MIMO SIGNALS AND SYSTEMS

Coded Modulation Systems
John B. Anderson and Arne Svensson

Communication System Design Using DSP Algorithms: With Laboratory Experiments for the TMS320C6701 and TMS320C6711
Steven A. Tretter

Interference Avoidance Methods for Wireless Systems
Dimitrie C. Popescu and Christopher Rose

MIMO Signals and Systems
Horst J. Bessai

Performance Analysis and Modeling of Digital Transmission Systems
William Turin

Stochastic Image Processing
Chee Sun Won and Robert M. Gray

Wireless Communications Systems and Networks
Mohsen Guizani

MIMO SIGNALS AND SYSTEMS

Horst J. Bessai
University of Siegen
Siegen, Germany

Springer

Library of Congress Cataloging-in-Publication Data

Bessai, Horst J.
MIMO signals and systems / by Horst J. Bessai.
p. cm. -- (Information technology--transmission, processing, and storage)
Includes bibliographical references and index.
ISBN 0-387-23488-8
1. Signal processing--Mathematics. 2. MIMO systems. 3. Wireless communication systems. I. Title. II. Series.

TK5102.9.B47 2005
621.382'2--dc22

2004060675

ISBN- 1-4899-9367-3

9 8 7 6 5 4 3 2 1

springeronline.com

Dedicated to my wife Ellen
and my son Jan
for their love, understanding, and support.

PREFACE

This text evolved from notes used to teach two-semester courses on multi-port signals and systems theory and vector-valued signal transmission to third-year electrical and computer engineering students. It is also based on the author's tutorial courses on the subject presented to practicing engineers in industry. The primary motivation has been to familiarize the reader with the essential tools and methods used to describe the dynamic behavior of electrical multiple-input multiple-output (MIMO) systems. The book shall provide a basic understanding of the fundamentals, implementation, and applications of MIMO techniques. For easier comprehension, these techniques, in conjunction with several "classic" algorithms, are illustrated by means of numerous worked examples. MATLAB, a matrix-oriented commercial software package with user-friendly interfaces and excellent graphics support, was chosen to perform numerical analyses. MATLAB is very easy to learn and *de facto* a worldwide standard programming language in universities and industry. End-of chapter problems are added to provide additional training opportunities and to reinforce the knowledge gained.

Over the last decade, spurred by the invention of a series of fundamentally new wireless transmission concepts, MIMO theory has been transformed into one of the most vibrant and active research areas. Communications engineers continue to produce – at an unprecedented high speed – more accurate radio channel models. Spectral efficiencies of actually working systems are reported as high as 20 bits/s/Hz. Information theorists are eager to find more accurate formulas describing capacity bounds for communication systems with multiple transmit and/or receive antennas.

Large teams of scientists are working hard on the development of novel space-time codes. Others try to design miniaturized antenna arrays.

Even several of the 34 "state-of-the-art" articles published in two special issues (April + June 2003) of the *IEEE Journal on Selected Areas in Communications* are, to some extent, outdated because of more recent extensions and technical improvements. New results of ongoing MIMO research activities appear almost every week. Today, even an initial literature search into the "basics" of MIMO systems yields a minimum of 2000 articles that appeared in journals, magazines, and conference papers.

A person with no previous training or experience in the specific fields of electromagnetics, radio technology, wireless transmission systems, coding and digital modulation theory, etc., is in danger of getting lost right at the start of a MIMO project. It is the purpose of this text to take away at least some (not all!) of the load from my students and from the interested readers.

Despite the large quantity of excellent publications, there appears to be a rift between sophisticated MIMO theory and practical applications. A complaint often heard from students is that, in most professional journals and textbooks, mathematics multiplies beyond necessity. I have, therefore, made all efforts to minimize mathematical prerequisites as well as the number of extra computational steps to be taken while reading through the text. Moreover, by omitting more than 50 percent of my original manuscript, I tried hard to avoid all details that were considered as either too specialized or in danger of being outdated by the day of the book's publication. Of course, this comes with the risk of not capturing essential subjects in their full breadth...

The book starts out in chapter 1 with a comprehensive review of linear time invariant multiport systems. Signals of the type $\underline{s}(\mathbf{p}, t)$, where $\mathbf{p}$ is a geometric position vector and t is the continuous time variable, are then considered in chapter 2. It shall serve as an introduction into physics of the generation and propagation of electromagnetic waves through linear media. The third chapter covers some of the basics needed to design antennas and to understand the mechanisms of radiation of electromagnetic waves. Chapter 4 deals with signal space concepts. Tools and algorithms are presented to perform various standard signal processing tasks such as projection, orthogonalization, orthonormalization, and QR-decomposition of discrete-time signals. Simple numerical examples are frequently included in the text to demonstrate how the algorithms actually work. Finally, in chapter 5, a systematic assortment of MIMO channel problems are listed and discussed. Several of these problems are numerically solved by means of standard tools known from linear algebra (vector calculus, matrix inverse and pseudoinverse) and probability theory. Others require more sophisticated techniques such as maximum likelihood symbol detection. As an example of the utilization of space-time codes, the Alamouti scheme is explained for a scenario with two transmit antennas and one receive antenna. All steps are demonstrated for the transmission of a QPSK-modulated stream of information-carrying data symbols. A MATLAB program is provided with an implementation of both the transmit and the receive side of a QPSK-modulated Alamouti scheme. Extensions towards orthogonal designs with both real and complex-valued space-time code matrices are straight forward. Various known STC code matrices are listed in appendix C.

The first two appendices (A and B) include collections of formulas needed in vector analysis (appendix A) and matrix algebra (appendix B).

It is my aim to provide the students with some background in the areas of MIMO signal and system theory. I should emphasize that it is possible to teach the whole material without any significant omissions in a two-semester course. Several extra topics of interest were intentionally left out and are to be covered in the seminars and/or ongoing project works leading to a series of master and doctoral theses. Topics not covered include dynamic channel modeling, equalization of MIMO radio channels, carrier and clock synchronization techniques, antenna arrays and beamforming. New results, corrections as well as updates and additions to the MATLAB program files will be made available on my website at http://bessai.coolworld.de

Horst J. Bessai

University of Siegen, Germany
Siegen, June 2004.

Acknowledgment

I am indebted to Ana Bozicevic and Alexander N. Greene of Kluwer Academic Publishers for their encouragement and help.

Contents

ADOPTED NOTATIONS

The following conventions are used throughout the text.

a	Italic letters are used to denote real-valued scalar variables.
$\underline{a}$	Underlined italic letters denote complex variables.
$\underline{a}^*$	Complex conjugate of complex variable $\underline{a}$.
$j=\sqrt{-1}$	Complex variable.
$Re\{\underline{a}\}$	Real part of complex variable $\underline{a}$.
$Im\{\underline{a}\}$	Imaginary part of complex variable $\underline{a}$.
$\lvert\underline{a}\rvert$	Symbol $\lvert\ \rvert$ is used to denote the absolute value or magnitude of the complex scalar variable $\underline{a}$ enclosed within.
$\angle\underline{a}$	Phase angle of complex variable $\underline{a}$.
$\mathbf{a}$	Column vectors are denoted by boldfaced lowercase letters.
$\mathbf{A}$	Matrices are denoted by boldfaced uppercase letters. Boldfaced uppercase letters are also used to denote vectors of electromagnetic fields such as **E**, **H**, etc.
$\hat{a}, \hat{\mathbf{a}}, \hat{\mathbf{A}}$	Hat (^) placed over the symbol denotes an estimate of a scalar, vector, or matrix. Symbol hat (^) is also used to designate the amplitude of a complex-valued signal such as $\underline{s}(t)=\hat{s}\,exp(j\omega t)$.
$\lvert\mathbf{a}\rvert$	Magnitude or length of vector **a**.
$\mathbf{a}^T, \mathbf{A}^T$	Transpose of vector **a** or matrix **A**, respectively.
$\mathbf{a}^H, \mathbf{A}^H$	Hermitian (complex conjugate and transposed) of vector **a** or matrix **A**, respectively.

$\mathbf{A}^{-1}$	Inverse of matrix **A**.
$\mathbf{A}^{+}$	Pseudoinverse of matrix **A** where **A** is not necessarily a square matrix.
$\lVert\mathbf{A}\rVert_F$	Frobenius norm of matrix **A**.
$\angle(\mathbf{a},\mathbf{b})$	Phase angle between vectors **a** and **b**.
►◄	Designation of the end of examples.

Note: All vectors are kept in column form.

Chapter 1

REVIEW OF LINEAR TIME-INVARIANT (LTI) MULTI-PORT SYSTEM THEORY

1.1 INTRODUCTION

In this chapter, we provide a concise introduction to the theory underlying the transmission of signals over channels with multiple input ports and multiple output ports, so-called MIMO channels. To investigate the properties of these channels, it is necessary to model them based upon their electrical characteristics, i.e., relations between sets of voltages, currents, and powers measured at their accessible nodes. Therefore, the focus of the following Section 1.2 is on matrix calculus as applied to loop and nodal analysis of electrical networks. We shall see how an arbitrarily complex network with P non-accessible interior nodes, N input ports, and M outputs can be described by its admittance matrix. Once we have such a matrix available, it is possible to reduce the number of interior nodes. A node reduction algorithm (NRA) is presented, and equivalent two-port circuits are derived that characterize the network's behavior between any one of the inputs and a selected output port.

Although the admittance matrix approach yields very compact mathematical descriptions of multi-ports, it does not adequately cover the idea of waves traveling through a transmission medium at high frequencies. When dealing with a multi-port's incident waves, reflected and transmitted waves, we should change the description method and switch to the then more appropriate concept of scattering or S parameters. Therefore, in Section 1.3, both normalized and generalized scattering matrix models are introduced and applied to multi-port networks. To be as flexible as possible, we assume that every port may have its own complex termination or excitation, which leads us to the notion of generalized scattering parameters. Thus, possible mismatching effects between a port's input impedance at a given frequency and the characteristic impedance of the termination can be properly evaluated. Moreover, we shall see how multi-port networks with known S matrices can be cascaded, with a set of output ports of the first network connected back-to-back to the same number of input ports of the second network.

Then, in Section 1.4, we shall have a closer look at the essential system theoretical properties of MIMO channels. We go back to the standard classification of time-continuous and time-discrete systems and slightly broaden the well-known definitions with respect to multiple system inputs and outputs. Throughout the remainder of this book, our further terminology and analyses will be based upon these definitions. The versatile concept of state variables will be explained and applied to multi-ports.

Though several examples and supporting MATLAB® programs are provided with the main text, it is recommended that the reader go through the problems in Section 1.5.

1.2 A MATRIX-BASED VIEW OF ELECTRICAL MULTI-PORT NETWORKS

We start our brief review of linear time-invariant systems with a simple "black box" description of a **two-port network**. It has two nodes at both its input and output as shown in Figure 1-1. Note that, by definition, throughout this text we shall use symmetrical reference currents. Symbolically, all current arrows point *into* the network.

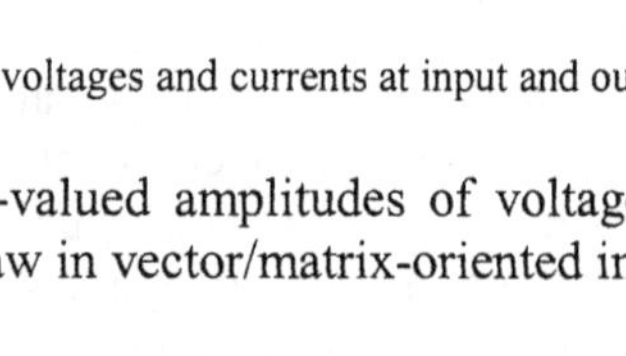

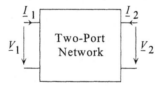

Figure 1-1. Definition of voltages and currents at input and output of a two-port network.

Assuming complex-valued amplitudes of voltages $\underline{V}$ and currents $\underline{I}$, we should rewrite Ohm's law in vector/matrix-oriented impedance notation as

$$\begin{pmatrix} \underline{V}_1 \\ \underline{V}_2 \end{pmatrix} = \begin{pmatrix} \underline{Z}_{1,1} & \underline{Z}_{1,2} \\ \underline{Z}_{2,1} & \underline{Z}_{2,2} \end{pmatrix} \cdot \begin{pmatrix} \underline{I}_1 \\ \underline{I}_2 \end{pmatrix}, \tag{1.1}$$

or, equivalently, using admittances,

$$\begin{pmatrix} \underline{I}_1 \\ \underline{I}_2 \end{pmatrix} = \begin{pmatrix} \underline{Y}_{1,1} & \underline{Y}_{1,2} \\ \underline{Y}_{2,1} & \underline{Y}_{2,2} \end{pmatrix} \cdot \begin{pmatrix} \underline{V}_1 \\ \underline{V}_2 \end{pmatrix}. \tag{1.2}$$

Obviously, the electrical properties of this network can be completely described by its external voltages and currents. This is done either through four complex-valued impedances or by means of admittances, arranged in the form of (2×2) matrices **Z** or **Y**, respectively. More generally, it should be possible to describe every multi-port network with N input ports and M output ports by an $(N+M)\times(N+M)$ **impedance matrix**

$$\mathbf{Z}=\begin{pmatrix} \underline{Z}_{1,1} & \underline{Z}_{1,2} & \cdots & \underline{Z}_{1,(N+M)} \\ \underline{Z}_{2,1} & \underline{Z}_{2,2} & \cdots & \underline{Z}_{2,(N+M)} \\ \vdots & \vdots & \ddots & \vdots \\ \underline{Z}_{(N+M),1} & \underline{Z}_{(N+M),2} & \cdots & \underline{Z}_{(N+M),(N+M)} \end{pmatrix}, \tag{1.3}$$

or in **admittance matrix** form

$$\mathbf{Y}=\begin{pmatrix} \underline{Y}_{1,1} & \underline{Y}_{1,2} & \cdots & \underline{Y}_{1,(N+M)} \\ \underline{Y}_{2,1} & \underline{Y}_{2,2} & \cdots & \underline{Y}_{2,(N+M)} \\ \vdots & \vdots & \ddots & \vdots \\ \underline{Y}_{(N+M),1} & \underline{Y}_{(N+M),2} & \cdots & \underline{Y}_{(N+M),(N+M)} \end{pmatrix}. \tag{1.4}$$

Input ports carry numbers 1 through N and, starting with number $N+1$, we continue counting the M output ports from $N+1$ to $N+M$. Voltages and currents are assigned to these ports as depicted in Figure 1-2.

Of course, the question arises how to calculate the matrix elements $\underline{Z}_{i,j}$ and $\underline{Y}_{i,j}$ for a given (N, M) network. Both indices, i and j, range from 1 to $N+M$, and it should be mentioned that **Z** and **Y** are always square $(N+M)\times(N+M)$ matrices.

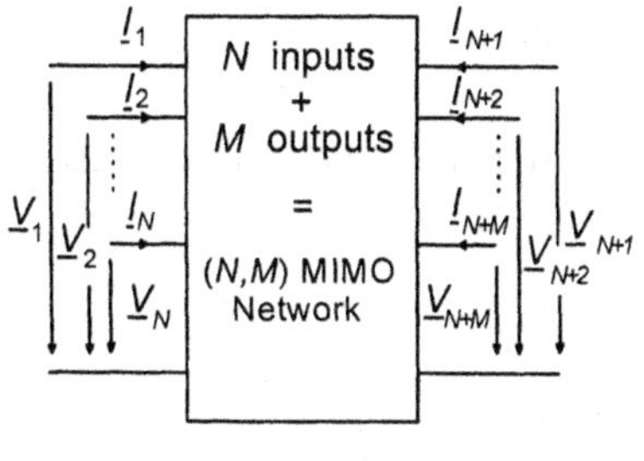

Figure 1-2. Voltages and currents at the ports of (N, M) MIMO system.

By rewriting $\mathbf{V} = \mathbf{Z}\cdot\mathbf{I}$ as a set of $N+M$ linear equations, we obtain a simple instruction for the calculation of all unknown impedance matrix elements $\underline{Z}_{i,j}$. We measure the open-circuit voltage $\underline{V}_i$ at port i while port j is driven by current $\underline{I}_j$. All other ports with indices $k \neq j$ are open-circuited (that is, $\underline{I}_k = 0$ for $k \neq j$). By calculating the ratios

$$\underline{Z}_{i,j} = \frac{\underline{V}_i}{\underline{I}_j}\Bigg|_{\underline{I}_k = 0}, \quad \text{for } k \neq j \tag{1.5}$$

we get, in the most general case, $(N+M)^2$ complex-valued elements of matrix $\mathbf{Z}$. Thus, the total number of degrees of freedom amounts to $2(N+M)^2$. Fortunately, in practice, there are many situations where $\mathbf{Z}$ becomes either a symmetrical matrix (reciprocal networks with $\underline{Z}_{i,j} = \underline{Z}_{j,i}$ or, in case of loss-less networks, all $\underline{Z}_{i,j}$ elements are purely imaginary with $\mathrm{Re}\{\underline{Z}_{i,j}\} = 0$.

From $\mathbf{I} = \mathbf{Y}\cdot\mathbf{V}$ we can find the complex-valued admittance elements, $\underline{Y}_{i,j}$, as

$$\underline{Y}_{i,j} = \frac{\underline{I}_i}{\underline{V}_j}\Bigg|_{\underline{V}_k = 0}, \quad \text{for } k \neq j \,. \tag{1.6}$$

Here one has to measure the driving voltage at port j and the short-circuit current at port i with all ports other than j short-circuited. Again, $\underline{Y}_{i,j} = \underline{Y}_{j,i}$ holds for reciprocal networks, and pure imaginary $\underline{Y}_{i,j}$'s exist if the network is lossless.

EXAMPLE 1-1: Consider the two-port networks shown in Figure 1-3. For the T-shaped circuit (*a*) application of (1.1) and (1.5) yields the impedance matrix

$$\mathbf{Z}_T = \begin{pmatrix} \underline{Z}_{T1} + \underline{Z}_{T3} & \underline{Z}_{T3} \\ \underline{Z}_{T3} & \underline{Z}_{T2} + \underline{Z}_{T3} \end{pmatrix}. \tag{1.7}$$

The admittance matrix of the π-shaped circuit (*b*) can be calculated using (1.2) and (1.6). We get

$$\mathbf{Y}_\pi = \begin{pmatrix} \underline{Y}_{\pi 1} + \underline{Y}_{\pi 3} & -\underline{Y}_{\pi 3} \\ -\underline{Y}_{\pi 3} & \underline{Y}_{\pi 2} + \underline{Y}_{\pi 3} \end{pmatrix}. \tag{1.8}$$

Note that networks (*a*) and (*b*) exhibit identical external properties if $\mathbf{Z}_T = \mathbf{Z}_\pi = \mathbf{Y}_\pi^{-1}$.

▶◀

Useful conversion formulas from π to T (and T to π, respectively) are displayed in Table 1-1.

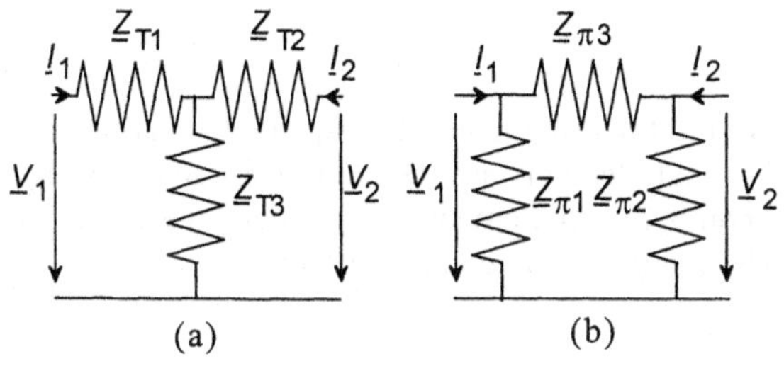

Figure 1-3. Two-port T-network (*a*) and π-network (*b*).

Table 1-1. Elements of equivalent T- and π-shaped two-port networks.

Conversion: T to π	Conversion: π to T
$\underline{Z}_{\pi 1} = \frac{\underline{Z}_{T1}\underline{Z}_{T2} + \underline{Z}_{T1}\underline{Z}_{T3} + \underline{Z}_{T2}\underline{Z}_{T3}}{\underline{Z}_{T2}}$	$\underline{Z}_{T1} = \frac{\underline{Y}_{\pi 2}}{\underline{Y}_{\pi 1}\underline{Y}_{\pi 2} + \underline{Y}_{\pi 1}\underline{Y}_{\pi 3} + \underline{Y}_{\pi 2}\underline{Y}_{\pi 3}}$
$\underline{Z}_{\pi 2} = \frac{\underline{Z}_{T1}\underline{Z}_{T2} + \underline{Z}_{T1}\underline{Z}_{T3} + \underline{Z}_{T2}\underline{Z}_{T3}}{\underline{Z}_{T1}}$	$\underline{Z}_{T2} = \frac{\underline{Y}_{\pi 1}}{\underline{Y}_{\pi 1}\underline{Y}_{\pi 2} + \underline{Y}_{\pi 1}\underline{Y}_{\pi 3} + \underline{Y}_{\pi 2}\underline{Y}_{\pi 3}}$
$\underline{Z}_{\pi 3} = \frac{\underline{Z}_{T1}\underline{Z}_{T2} + \underline{Z}_{T1}\underline{Z}_{T3} + \underline{Z}_{T2}\underline{Z}_{T3}}{\underline{Z}_{T3}}$	$\underline{Z}_{T3} = \frac{\underline{Y}_{\pi 3}}{\underline{Y}_{\pi 1}\underline{Y}_{\pi 2} + \underline{Y}_{\pi 1}\underline{Y}_{\pi 3} + \underline{Y}_{\pi 2}\underline{Y}_{\pi 3}}$

Next, we attempt to describe arbitrarily complex (N, M) multi-ports with interior (non-accessible) and external nodes. The following **Node Reduction Algorithm (NRA)** considers the particular structure of the network and yields a **reduced admittance matrix**, denoted by $\tilde{\mathbf{Y}}$.

• **Step 1**: Count the network's number of **internal** and **exterior nodes** (say P, excluding ground = node 0), and assign an index to each node. . . .

• **Step 2**: Write down the network's square $P{\times}P$ admittance matrix, **Y**, such that diagonal elements are the sum of all admittances connected to node i ,and non-diagonal elements are the negative values of admittances between nodes i and j (direct connections only!).

• **Step 3**: Re-arrange lines and columns of admittance matrix **Y** such that the first two rows and columns (upper left hand corner of **Y**) correspond to the desired input and output of the two-port network.

• **Step 4**: Obtain the reduced 2×2 version of the admittance matrix, called $\tilde{\mathbf{Y}}$, by taking four **principal submatrices** and calculating

$$\tilde{\mathbf{Y}} = \mathbf{Y}_{1,2|1,2} - \mathbf{Y}_{1,2,|3,\cdots,P} \cdot \mathbf{Y}_{3,\cdots,P|3,\cdots,P}^{-1} \cdot \mathbf{Y}_{3,\cdots,P|1,2} \tag{1.9}$$

where indices on the left of the vertical bar represent lines of the original matrix, **Y**, and those on the right of the bar stand for columns to be cut out of **Y**.

Note 1: Step 3 can be eliminated by assigning node indices such that node 1 is the two-port's input, and node 2 represents the output.

Note 2: If we wish to describe the network by only $Q > 2$ external ports (out of P), (1.9) can be easily modified to yield a $Q{\times}Q$ matrix, $\tilde{\mathbf{Y}}$, as follows:

$$\begin{pmatrix} \underline{I}_1 \\ \underline{I}_2 \\ \vdots \\ \underline{I}_Q \end{pmatrix} = \tilde{\mathbf{Y}} \cdot \begin{pmatrix} \underline{V}_1 \\ \underline{V}_2 \\ \vdots \\ \underline{V}_Q \end{pmatrix} \tag{1.10}$$

where

$$\tilde{\mathbf{Y}} = \mathbf{Y}_{1,\cdots,Q|1,\cdots,Q} - \mathbf{Y}_{1,\cdots,Q|Q+1,\cdots,N} \cdot \mathbf{Y}^{-1}_{Q+1,\cdots,N|Q+1,\cdots,N} \cdot \mathbf{Y}_{Q+1,\cdots,N|1,\cdots,Q} \cdot \quad (1.11)$$

Note 3: Three related MATLAB® programs (*m*-files) are available on the accompanying CD-ROM. Program "**nra.m**" is an implementation of the NRA algorithm. For a specified number of nodes (external + interior), *P*, the user is requested to enter the diagonal and off-diagonal admittance matrix elements. Then, after specifying the number of desired remaining nodes, Q, the elements of $Q \times Q$ matrix $\tilde{\mathbf{Y}}$ are calculated and displayed on the screen. If $Q = 2$ is chosen, the result will be an equivalent four-port representation of the original network. Programs "**mimo2t.m**" and "**mimo2pi.m**" are extended versions of "**nra.m**". The original multi-port network is transformed into an equivalent T- or π-circuit. The four two-port matrix elements are presented as complex numbers and, for a specified operating frequency, resistances and capacitances or inductances of the T- or π-circuit are computed.

The **voltage transfer function** (hence the index *V*) $\underline{H}_V$ is given by

$$\underline{H}_V = \frac{\underline{V}_2}{\underline{V}_1} = \frac{-\underline{\tilde{Y}}_{2,1}}{\underline{\tilde{Y}}_{2,2} + \underline{Y}_2} \quad (1.12)$$

where $\underline{Y}_2$ is the real or complex-valued load admittance on port 2 (output).

If that particular load admittance is zero, (1.12) reduces to the open-circuit voltage transfer function, $\underline{H}_V$, is to be distinguished from the **power transfer function**

$$\underline{H}_P = \frac{\underline{P}_1}{\underline{P}_2} = \frac{\underline{V}_2^2 \underline{Y}_2}{\underline{V}_1^2 \underline{Y}_{i,1}} = \frac{\underline{Y}_2}{\underline{Y}_{i,1}} \underline{H}_V^2 \quad (1.13)$$

where

$$\underline{Y}_{i,1} = \underline{\tilde{Y}}_{1,1} - \frac{\underline{\tilde{Y}}_{1,2} \underline{\tilde{Y}}_{2,1}}{\underline{\tilde{Y}}_{2,2} + \underline{Y}_2} \quad (1.14)$$

represents the input admittance at port 1. By inserting (1.14) into (1.13), we see that the power transfer function can be expressed in terms of the elements of 2×2 matrix $\tilde{\mathbf{Y}}$ and of the **output load admittance** $\underline{Y}_2$. It is thus given by

$$\underline{H}_P = \frac{\underline{P}_1}{\underline{P}_2} = \frac{\underline{Y}_2}{\underline{Y}_{i,1}} \underline{H}_V^2 = \frac{\underline{\widetilde{Y}}_{2,1}^2}{(1+\underline{\widetilde{Y}}_{2,2})\,(\det(\widetilde{\mathbf{Y}}) + \underline{\widetilde{Y}}_{1,1}\underline{Y}_2)}. \qquad (1.15)$$

EXAMPLE 1-2: Figure 1.4 shows a passive, hence reciprocal 4-port circuit. The network can be described in matrix notation by the voltages at its output nodes referenced to a single common reference node. These voltages are related to the currents entering the nodes by

$$\begin{pmatrix} \underline{I}_1 \\ \underline{I}_2 \\ \underline{I}_3 \\ \underline{I}_4 \end{pmatrix} = \mathbf{Y} \cdot \begin{pmatrix} \underline{U}_1 \\ \underline{U}_2 \\ \underline{U}_3 \\ \underline{U}_4 \end{pmatrix} \qquad (1.16)$$

where $\mathbf{Y}$ is a 4×4 matrix. We consider the resistors by their conductance $G = 1/R$. Then, after performing steps 1 and 2 of the NRA, we find

$$\mathbf{Y} = \begin{array}{c} \begin{matrix} 1 & 2 & 3 & 4 \end{matrix} \\ \begin{pmatrix} G & -G & 0 & 0 \\ -G & 2G + j\omega C & -G & 0 \\ 0 & -G & 2G + j\omega C & -G \\ 0 & 0 & -G & G \end{pmatrix} \begin{matrix} 1 \\ 2 \\ 3 \\ 4 \end{matrix} \end{array}. \qquad (1.17)$$

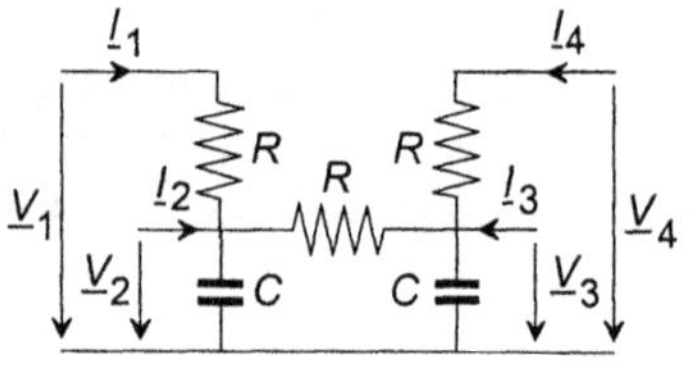

Figure 1-4. A simple passive four-port network

Now, assume that we consider **port 1** (= single input) and **port 3** (= single output) only. We rearrange the matrix lines and columns accordingly to get indices 1 and 3 in the upper left-hand corner of $\mathbf{Y}$. Thus, we have

$$\mathbf{Y} = \begin{array}{c} \begin{matrix} 1 & 3 & 2 & 4 \end{matrix} \\ \begin{pmatrix} G & 0 & -G & 0 \\ 0 & 2G + j\omega C & -G & -G \\ -G & -G & 2G + j\omega C & 0 \\ 0 & -G & 0 & G \end{pmatrix} \begin{matrix} 1 \\ 3 \\ 2 \\ 4 \end{matrix} \end{array}. \qquad (1.18)$$

Using (1.11), we calculate the reduced 2×2 matrix

$$\tilde{\mathbf{Y}} = \mathbf{Y}_{1,3|1,3} - \mathbf{Y}_{1,3|2,4} \cdot \mathbf{Y}_{2,4|2,4}^{-1} \cdot \mathbf{Y}_{2,4|1,3} \,. \tag{1.19}$$

By taking the sub-matrices in (1.18), we get

$$\tilde{\mathbf{Y}} = \begin{pmatrix} G & 0 \\ 0 & 2G + j\omega C \end{pmatrix} - \begin{pmatrix} -G & 0 \\ -G & -G \end{pmatrix} \cdot \begin{pmatrix} 2G + j\omega C & 0 \\ 0 & G \end{pmatrix}^{-1} \cdot \begin{pmatrix} -G & -G \\ 0 & -G \end{pmatrix}. \tag{1.20}$$

With a few matrix calculations and some reshuffling of terms we obtain

$$\tilde{\mathbf{Y}} = \frac{G^2}{2G + j\omega C} \begin{pmatrix} 1 + j\omega C/G & -1 \\ -1 & 1 - \left(\frac{\omega C}{G}\right)^2 + 3j\omega C/G \end{pmatrix}. \tag{1.21}$$

Finally, from (1.12) we know that then the open-circuit voltage transfer function is

$$\underline{H}_V = \frac{\underline{V}_3}{\underline{V}_1} = \frac{-\underline{\tilde{Y}}_{2,1}}{\underline{\tilde{Y}}_{2,2}} = \frac{1}{1 - \left(\frac{\omega C}{G}\right)^2 + 3j\omega C/G}. \tag{1.22}$$

►◄

So far, we assumed a common local reference node, or ground, indexed 0. There are, however, various applications where we don't have a common reference node. Also, in some cases, it may become necessary to interchange nodes. Then, to solve these problems, we define an **augmented admittance matrix**, $^{+}\mathbf{Y}$, that has an extra line, $P + 1$, and an additional column, $P + 1$, as follows:

$$^{+}\mathbf{Y} = \begin{pmatrix} \underline{Y}_{1,1} & \underline{Y}_{1,2} & \cdots & \underline{Y}_{1,P} & -\sum_{j=1}^{P} \underline{Y}_{1,j} \\ \underline{Y}_{2,1} & \underline{Y}_{2,2} & \cdots & \underline{Y}_{2,P} & -\sum_{j=1}^{P} \underline{Y}_{2,j} \\ \vdots & \vdots & \ddots & \vdots & \vdots \\ \underline{Y}_{P,1} & \underline{Y}_{P,2} & \cdots & \underline{Y}_{P,P} & -\sum_{j=1}^{P} \underline{Y}_{P,j} \\ -\sum_{i=1}^{P} \underline{Y}_{i,1} & -\sum_{i=1}^{P} \underline{Y}_{i,2} & \cdots & -\sum_{i=1}^{P} \underline{Y}_{i,P} & \sum_{i=1}^{P}\sum_{j=1}^{P} \underline{Y}_{i,j} \end{pmatrix}. \tag{1.23}$$

The ith element of the added column is the negative sum of all admittances in the ith line of the original matrix, $\mathbf{Y}$. For the jth element of the additional row vector, we sum up all elements of that row and enter the negative of that sum. Finally, the new diagonal element is the algebraic sum of all admittances in the non-augmented matrix, $\mathbf{Y}$. Note that, because of

$$det(^{+}\mathbf{Y}) = 0 \tag{1.24}$$

and $^{+}\mathbf{Z} = {}^{+}\mathbf{Y}^{-1}$ there exists no augmented impedance matrix. With the extra node, P+1, $^{+}\mathbf{Y}$ can be restructured with respect to different common reference node. For instance, the original output node 2 could become common node 0 ("ground") in the new circuit diagram.

An almost trivial exercise shall explain the procedures of matrix augmentation and interchange of ports.

EXAMPLE 1-3: Suppose we want to cut off the lower terminal of impedance $\underline{Z}_{T3}$ in Figure 1-3 (a) and make it the output terminal of a different T-network. Obviously, now the roles of $\underline{Z}_{T2}$ and $\underline{Z}_{T3}$ are interchanged. We call the original admittance matrix displayed in Figure 1-3 (a) $\mathbf{Y_T}^{(\mathbf{A})}$ and the new one $\mathbf{Y_T}^{(\mathbf{B})}$. Using (1.7), admittance matrix $\mathbf{Y_T}^{(\mathbf{A})}$ is given by

$$\mathbf{Y}_{\mathrm{T}}^{(\mathrm{A})} = \frac{1}{\underline{Z}_{T1}\underline{Z}_{T2}+\underline{Z}_{T1}\underline{Z}_{T3}+\underline{Z}_{T2}\underline{Z}_{T3}} \begin{array}{c} \begin{matrix} 1 & \qquad\qquad 2 \end{matrix} \\ \begin{pmatrix} \underline{Z}_{T2}+\underline{Z}_{T3} & -\underline{Z}_{T3} \\ -\underline{Z}_{T3} & \underline{Z}_{T1}+\underline{Z}_{T3} \end{pmatrix} \end{array} \begin{matrix} \\ 1 \\ 2 \end{matrix}. \tag{1.25}$$

Augmentation, according to (1.23), yields

$$\mathbf{Y}_{\mathrm{T}}^{(\mathrm{A})} = \frac{1}{\underline{Z}_{T1}\underline{Z}_{T2}+\underline{Z}_{T1}\underline{Z}_{T3}+\underline{Z}_{T2}\underline{Z}_{T3}} \begin{array}{c} \begin{matrix} 1 & \qquad 2 & \qquad 3 \end{matrix} \\ \begin{pmatrix} \underline{Z}_{T2}+\underline{Z}_{T3} & -\underline{Z}_{T3} & -\underline{Z}_{T2} \\ -\underline{Z}_{T3} & \underline{Z}_{T1}+\underline{Z}_{T3} & -\underline{Z}_{T1} \\ -\underline{Z}_{T2} & -\underline{Z}_{T1} & \underline{Z}_{T1}+\underline{Z}_{T2} \end{pmatrix} \end{array} \begin{matrix} \\ 1 \\ 2 \\ 3 \end{matrix} \tag{1.26}$$

By interchanging line and column 2 and 3, we make $\underline{Z}_{T3}$ the impedance connected to the output, while $\underline{Z}_{T2}$ is now connected to node 0, the common reference node. We omit line 3 and column 3 in $\mathbf{Y_T}^{(\mathbf{B})}$ and obtain

$$\mathbf{Y}_{\mathrm{T}}^{(\mathrm{B})} = \frac{1}{\underline{Z}_{T1}\underline{Z}_{T2}+\underline{Z}_{T1}\underline{Z}_{T3}+\underline{Z}_{T2}\underline{Z}_{T3}} \begin{pmatrix} \underline{Z}_{T2}+\underline{Z}_{T3} & -\underline{Z}_{T2} \\ -\underline{Z}_{T2} & \underline{Z}_{T1}+\underline{Z}_{T2} \end{pmatrix}. \tag{1.27}$$

The corresponding impedance matrix is the matrix inverse of $\mathbf{Y_T}^{(\mathbf{B})}$, namely

$$\mathbf{Z}_{\mathrm{T}}^{(\mathrm{B})} = \begin{pmatrix} \underline{Z}_{T1} + \underline{Z}_{T2} & \underline{Z}_{T2} \\ \underline{Z}_{T2} & \underline{Z}_{T2} + \underline{Z}_{T3} \end{pmatrix}. \tag{1.28}$$

▸◂

1.3 MICROWAVE MULTI-PORT NETWORKS WITH PORT EXCITATIONS

When dealing with electrical networks one should be able to consider port excitations at high frequencies, too. One of the salient modeling concepts is utilization of normalized scattering or S parameters. Especially, unlike in Y parameter analysis, there is no need to present a short-circuit to the ports. In S parameter analysis the network parameters are expressed in terms of **linear combinations** of voltages and currents at their terminals. Use is being made of relationships between incident and reflected waves normalized to a single characteristic impedance Z_0 (e.g., 50 Ω) or any set of possibly different characteristic impedances, considered in the form of a diagonal impedance matrix $\mathbf{Z}_0$. Suppose we have Q external ports (N of them are inputs and M are outputs). Then, as shown in Figure 1-5, the ith port (i = 1, 2, ... , Q) may be connected to a voltage source, modeled as an ideal voltage source $\underline{V}_{0i}$ in series with a real and positive characteristic source impedance Z_{0i}.

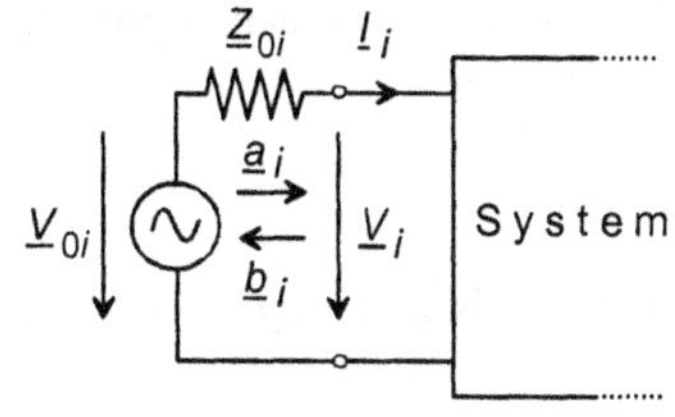

Figure 1-5. Modeling of excitation at ith port by scattering parameters.

Now, for the **incident voltage wave** $\underline{V}^+$ at the ith port, we define the complex-valued variable

$$\underline{a}_i = \frac{\underline{V}_i^+}{\sqrt{Z_{0i}}}. \tag{1.29}$$

Similarly, for the **reflected voltage wave** $\underline{V}^-$ at the ith port, we write

$$\underline{b}_i = \frac{\underline{V}_i^-}{\sqrt{Z_{0i}}}. \tag{1.30}$$

Alternatively, in terms of voltages and currents, the incident wave can be represented by

$$\underline{a}_i = \frac{\underline{V}_i + Z_{0i}\underline{I}_i}{2\sqrt{Z_{0i}}} = \frac{\underline{V}_{0i}}{2\sqrt{Z_{0i}}}, \tag{1.31}$$

and the reflected wave is considered by

$$\underline{b}_i = \frac{\underline{V}_i - Z_{0i}\underline{I}_i}{2\sqrt{Z_{0i}}}. \tag{1.32}$$

Note that, if the ith port is terminated by a real **characteristic impedance** Z_{0i} only, we have

$$\underline{V}_{0i} = 0,\ \underline{a}_i = 0\text{, and } \underline{b}_i = \underline{V}_i / \sqrt{Z_{0i}}. \tag{1.33}$$

Taking into account (1.31) and (1.32), we can express the voltage at port i as

$$\underline{V}_i = (\underline{a}_i + \underline{b}_i)\sqrt{Z_{0i}}, \tag{1.34}$$

and the current entering that port is given by

$$\underline{I}_i = \frac{\underline{a}_i - \underline{b}_i}{\sqrt{Z_{0i}}}. \tag{1.35}$$

Multiplication of $\underline{V}_i$ and $\underline{I}_i$ yields the amount of power, P_i, that is absorbed by the port's complex input impedance $\underline{Z}_{Ei} = \underline{V}_i / \underline{I}_i$. In terms of **scattering variables** and powers, we can write

$$P_i = \underline{V}_i \underline{I}_i = \underline{a}_i^2 - \underline{b}_i^2 = P_i^+ - P_i^-. \tag{1.36}$$

The square root of the ith source's characteristic impedance Z_{0i} normalizes

$\underline{a}_i$ and $\underline{b}_i$ with respect to the incident matched generator power

$$P^+ = |\underline{a}_i|^2 = \underline{a}_i \underline{a}_i^* \tag{1.37}$$

and to the reflected power

$$P^- = |\underline{b}_i|^2 = \underline{b}_i \underline{b}_i^* . \tag{1.38}$$

Hence, the dimensions of both $|\underline{a}_i|$ and $|\underline{b}_i|$ are square roots of power.

Furthermore, we use the ith port's complex input impedance $\underline{Z}_{Ei} = \underline{V}_i / \underline{I}_i$ and define the **reflection coefficient** as the ratio

$$\underline{r}_i = \frac{\underline{V}_i^-}{\underline{V}_i^+} = \frac{\underline{b}_i}{\underline{a}_i} = \frac{\underline{Z}_{Ei} - Z_{0i}}{\underline{Z}_{Ei} + Z_{0i}} . \tag{1.39}$$

Relating powers instead of voltages, we get the **return loss**

$$|\underline{r}_i|^2 = \frac{(\underline{V}_i^-)^2}{(\underline{V}_i^+)^2} = \frac{P_i^-}{P_i^+} . \tag{1.40}$$

It describes the amount of power reflected from port impedance $\underline{Z}_{Ei}$, and is usually expressed in decibels (dB) by

$$RL_i = -10\log(|\underline{r}_i|^2) = -20\log(|\underline{r}_i|) . \tag{1.41}$$

In a similar manner, the amount of incident power not reaching $\underline{Z}_{Ei}$ is defined by

$$\frac{P}{P^+} = 1 - |\underline{r}_i|^2 . \tag{1.42}$$

This power ratio is termed **mismatch loss**. Again, in decibels we have

$$ML_i = -10\log(1 - |\underline{r}_i|^2) . \tag{1.43}$$

Note that, in practice, there is always a need to define a so-called reference plane for each port with respect to incoming and outgoing waves. This is necessary to have a common node with respect to unit amplitude and zero phase.

Obviously, since $\underline{b}_i = \underline{r}_i \cdot \underline{a}_i$ and $i = 1, 2, \ldots Q$, for a Q-port system as shown in Figure 1-6, we should rewrite the **normalized scattering variables** in vector form as follows

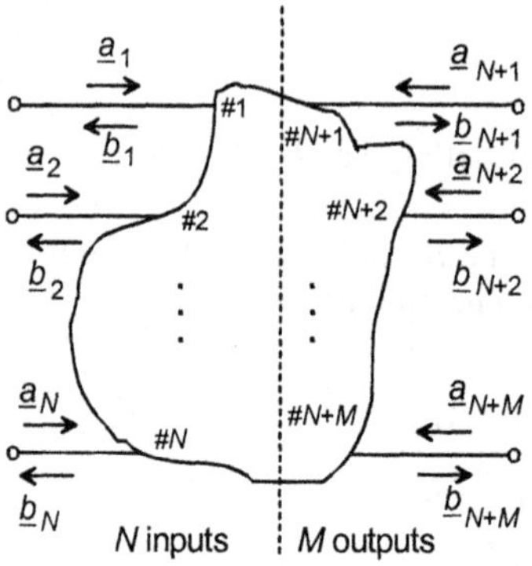

Figure 1-6. Definition of scattering variables for a Q-port network with N input terminals and M outputs. (Total number of ports is $Q = N + M$).

$$\mathbf{a} = (\underline{a}_1, \underline{a}_2, \cdots, \underline{a}_Q)^T \tag{1.44}$$

and

$$\mathbf{b} = (\underline{b}_1, \underline{b}_2, \cdots, \underline{b}_Q)^T . \tag{1.45}$$

Then, wave vectors **a** and **b** are related by a $Q \times Q$ **scattering matrix**, **S**, whose elements $\underline{S}_{i,j}$ are, in the most general case, complex-valued frequency-dependent variables.

In a more compact form, we write

$$\mathbf{b} = \mathbf{S} \cdot \mathbf{a} \tag{1.46}$$

where the ith diagonal element of scattering matrix **S** is obtained by the ratio of scattering variables at the same port, i.e.,

$$\underline{S}_{i,i} = \frac{\underline{b}_i}{\underline{a}_i}\bigg|_{\underline{a}_k=0,\text{ for all } k\neq i} \tag{1.47}$$

We notice that any one of these S parameters is identical with the ith reflection coefficients, $\underline{r}_i$, as defined in (1.39).

Furthermore, the off-diagonal elements of matrix **S** are given by the ratios

$$\underline{S}_{i,j} = \frac{\underline{b}_i}{\underline{a}_j}\bigg|_{\underline{a}_k=0,\text{ for all } k\neq j} \cdot \tag{1.48}$$

$|\underline{S}_{i,j}|^2$ is termed ***forward*** **insertion power gain** from port j to port i. With an interchange of indices i and j, we get the ***reverse*** **power gain** $|\underline{S}_{j,i}|^2$ which describes power transfer in the opposite direction, i.e., from port i to port j.

Note that the reflection coefficient at port i is equal to $\underline{S}_{i,i}$ only if all other ports $k \neq i$ are terminated with matched impedances. Another important point to remember is that the transmission coefficient from port j to port i is equal to $\underline{S}_{i,j}$ only if all other ports are matched.

By rewriting (1.34) and (1.35) in column vector form for a Q-port network, the **scattering variables** are obtained in the form of

$$\mathbf{a} = \frac{1}{2}\mathbf{D_0}\cdot(\mathbf{V}+\mathbf{Z_0}\cdot\mathbf{I}) = \frac{1}{2}\mathbf{D_0}\cdot\mathbf{V_0} \tag{1.49}$$

and

$$\mathbf{b} = \frac{1}{2}\mathbf{D_0}\cdot(\mathbf{V}-\mathbf{Z_0}\cdot\mathbf{I}) \tag{1.50}$$

where $\mathbf{D_0}$, $\mathbf{Z_0}$, and $\mathbf{V_0}$ are $Q\times Q$ diagonal matrices. They are defined as

$$\mathbf{D_0} = \begin{pmatrix} 1/\sqrt{Z_{01}} & 0 & \cdots & 0 \\ 0 & 1/\sqrt{Z_{02}} & \cdots & 0 \\ \vdots & \vdots & \ddots & \\ 0 & 0 & \cdots & 1/\sqrt{Z_{0Q}} \end{pmatrix}, \tag{1.51}$$

$$\mathbf{Z_0} = \begin{pmatrix} Z_{01} & 0 & \cdots & 0 \\ 0 & Z_{02} & \cdots & 0 \\ \vdots & \vdots & \ddots & \vdots \\ 0 & 0 & \cdots & Z_{0Q} \end{pmatrix}, \tag{1.52}$$

and, finally, for the source voltages we write

$$\mathbf{V_0} = \begin{pmatrix} V_{01} & 0 & \cdots & 0 \\ 0 & V_{02} & \cdots & 0 \\ \vdots & \vdots & \ddots & \vdots \\ 0 & 0 & \cdots & V_{0Q} \end{pmatrix}. \tag{1.53}$$

If a common real-valued **reference impedance** $Z_{01} = Z_{02} = \cdots = Z_{0Q} = Z_0$ (e.g., 50 Ω) is chosen, (1.51) and (1.52) become $Q \times Q$ identity (unity diagonal) matrices, **U**, multiplied by scalars $1/\sqrt{Z_0}$ and Z_0, respectively. (*Note*: To avoid a mix-up of variables between current vectors and identity matrices, we shall denote, as an exception, the identity matrix by a boldfaced uppercase letter **U**.) Hence, in cases with unequal reference impedances, we have

$$\mathbf{D_0} = \frac{1}{\sqrt{Z_0}} \mathbf{U} \tag{1.54}$$

and

$$\mathbf{Z_0} = Z_0 \, \mathbf{U}. \tag{1.55}$$

From $\mathbf{b} = \mathbf{S} \cdot \mathbf{a}$ and $\mathbf{V} = \mathbf{Z} \cdot \mathbf{I}$, by taking into account (1.49) and (1.50) and some matrix calculus, we obtain the following conversion formulas

$$\mathbf{S} = \mathbf{D_0} \cdot (\mathbf{Z} - \mathbf{Z_0}) \cdot (\mathbf{Z} + \mathbf{Z_0})^{-1} \cdot \mathbf{D_0^{-1}}, \tag{1.56}$$

and

$$\mathbf{S} = \mathbf{D_0} \cdot (\mathbf{U} - \mathbf{Y_0^{-1}} \cdot \mathbf{Y}) \cdot (\mathbf{U} + \mathbf{Y_0^{-1}} \cdot \mathbf{Y})^{-1} \cdot \mathbf{D_0^{-1}}. \tag{1.57}$$

Conversely, solving for matrices **Z** and **Y**, we find

$$\mathbf{Z} = \mathbf{D}_0^{-1} \cdot (\mathbf{U} - \mathbf{S})^{-1} \cdot (\mathbf{U} + \mathbf{S}) \cdot \mathbf{D}_0 \cdot \mathbf{Z}_0 \tag{1.58}$$

and

$$\mathbf{Y} = \mathbf{Y}_0 \cdot \mathbf{D}_0^{-1} \cdot (\mathbf{U} + \mathbf{S})^{-1} \cdot (\mathbf{U} - \mathbf{S}) \cdot \mathbf{D}_0 . \tag{1.59}$$

EXAMPLE 1-4: We consider the simple two-port network shown in Figure 1-6 and choose resistances $Z_0 = 50\ \Omega$, $Z_1 = 100\ \Omega$, and $Z_2 = 25\ \Omega$. Then, the input resistance at port 1 becomes

$$Z_{E1} = Z_1 + \left(Z_2 \| Z_0\right) = 116.67\,\Omega , \tag{1.60}$$

and the input resistance at port 2 is

$$Z_{E2} = \left(Z_0 + Z_1\right) \| Z_2 = 21.43\,\Omega . \tag{1.61}$$

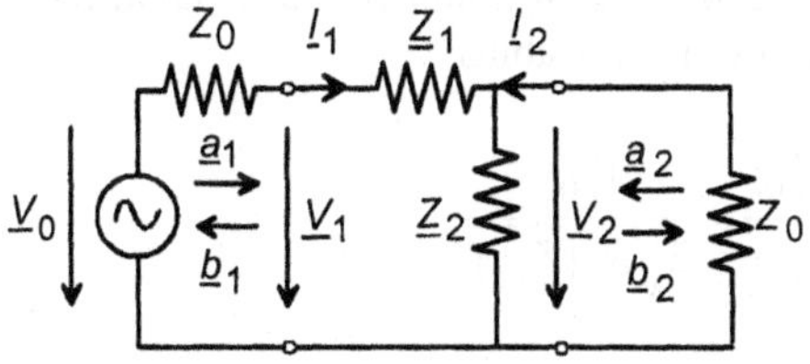

Figure 1-7. L-shaped two-port network of example 1-4.

The diagonal elements (**reflection coefficients**) of the network's scattering matrix **S** can be calculated as

$$\underline{S}_{11} = \left.\frac{\underline{a}_1}{\underline{b}_1}\right|_{\underline{a}_2=0} = \frac{Z_{E1} - Z_0}{Z_{E1} + Z_0} = 0.4 , \tag{1.62}$$

and for port 2 we obtain

$$\underline{S}_{22} = \left.\frac{\underline{a}_2}{\underline{b}_2}\right|_{\underline{a}_1=0} = \frac{Z_{E2} - Z_0}{Z_{E2} + Z_0} = -0.4 . \tag{1.63}$$

The non-diagonal elements are

$$\underline{S}_{12} = \frac{\underline{b}_1}{\underline{a}_2}\bigg|_{\underline{a}_1=0} = \frac{\underline{V}_1}{\underline{V}_2^+} = \\ = \frac{\underline{V}_2 2\sqrt{Z_0}}{\sqrt{Z_0}\,\underline{V}_{02}} = 2\frac{Z_2\|(Z_1+Z_0)}{(Z_2\|(Z_1+Z_0))+Z_0}\cdot\frac{Z_0}{(Z_1+Z_0)} = 0.1905 \tag{1.64}$$

and

$$\underline{S}_{21} = \frac{\underline{b}_2}{\underline{a}_1}\bigg|_{\underline{a}_2=0} = \frac{\underline{V}_2}{\underline{V}_1^+} = \\ = \frac{\underline{V}_2 2\sqrt{Z_0}}{\sqrt{Z_0}\,\underline{V}_{01}} = 2\frac{Z_2\|Z_0}{(Z_2\|Z_0)+(Z_0+Z_1)} = 0.2\,. \tag{1.65}$$

►◄

In the analysis of practical microwave systems, it may become necessary to consider **complex Thevenin impedances** (sources or loads), $\underline{Z}_{0i}$, at the ith port instead of pure resistive elements, as in the preceding network analysis. This leads us to the definition of **generalized *S* parameters**. They are based upon generalized scattering variables

$$\underline{a}'_i = \frac{\underline{V}_i + \underline{I}_i\underline{Z}_{0i}}{2\sqrt{Re\{\underline{Z}_{0i}\}}} = \frac{\underline{V}_{0i}}{2\sqrt{Re\{\underline{Z}_{0i}\}}} \tag{1.66}$$

and

$$\underline{b}'_i = \frac{\underline{V}_i - \underline{I}_i\underline{Z}^*_{0i}}{2\sqrt{Re\{\underline{Z}_{0i}\}}} \tag{1.67}$$

where $Re\{...\}$ denotes the real part of a complex variable. The power available from a voltage source connected to port i is, analogous to (1.36),

$$P_i^+ = |\underline{a}'_i|^2 = \underline{a}'_i\,\underline{a}'^*_i = \frac{\underline{V}_{0i}^2}{4\,Re\{\underline{Z}_{0i}\}}\,. \tag{1.68}$$

Furthermore, the power absorbed at the ith port is

$$
\begin{aligned}
P_i &= Re\{\underline{V}_i^* \underline{I}_i\} \\
&= |\underline{a}'_i|^2 - |\underline{b}'_i|^2 = P_i^+ - P_i^- \\
&= \frac{(\underline{Z}_{0i} + \underline{Z}_{0i}^*)(\underline{V}_i \underline{I}_i^* + \underline{V}_i^* \underline{I}_i)}{4\, Re\{\underline{Z}_{0i}\}} \\
&= P_i^+ (1 - |\underline{r}'_i|^2)
\end{aligned} \tag{1.69}
$$

where, similar to (1.38), we have

$$
\underline{r}'_i = \frac{\underline{b}'_i}{\underline{a}'_i} = \frac{\underline{V}_{0i} - \underline{I}_i \underline{Z}_{0i}^*}{\underline{V}_{0i} + \underline{I}_i \underline{Z}_{0i}} = \frac{\underline{Z}_{Ei} - \underline{Z}_{0i}^*}{\underline{Z}_{Ei} + \underline{Z}_{0i}} \tag{1.70}
$$

denotes the ith **generalized reflection coefficient**. In matrix notation, (1.46) is now written in terms of the generalized scattering variables, **a'** and **b'**, and the generalized scattering matrix, **S'**, as follows:

$$
\mathbf{b}' = \mathbf{S}' \cdot \mathbf{a}' . \tag{1.71}
$$

The two vectors of **generalized scattering variables** are defined by

$$
\mathbf{a}' = \frac{1}{2} \mathbf{D_0} \cdot (\mathbf{V} + \mathbf{Z_0} \cdot \mathbf{I}) \tag{1.72}
$$

and

$$
\mathbf{b}' = \frac{1}{2} \mathbf{D_0} \cdot (\mathbf{V} - \mathbf{Z_0^*} \cdot \mathbf{I}) . \tag{1.73}
$$

Note the complex conjugate elements of the reference impedance matrix $\mathbf{Z_0}$ in (1.73). In the $Q \times Q$ diagonal matrix $\mathbf{D_0}$ we consider *only the real part* of the complex reference impedances $\underline{Z}_{0i}$. Therefore, we take into account

$$\mathbf{D_0} = \begin{pmatrix} 1/\sqrt{\mathrm{Re}\{\underline{Z}_{01}\}} & 0 & \cdots & 0 \\ 0 & 1/\sqrt{\mathrm{Re}\{\underline{Z}_{02}\}} & \cdots & 0 \\ \vdots & \vdots & \ddots & \vdots \\ 0 & 0 & \cdots & 1/\sqrt{\mathrm{Re}\{\underline{Z}_{0Q}\}} \end{pmatrix}. \tag{1.74}$$

The diagonal elements of **S'** are given by

$$\begin{aligned} \underline{S}'_{i,i} &= \left.\frac{\underline{b}'_i}{\underline{a}'_i}\right|_{\underline{a}'_k=0 \quad \text{for all } k\neq i} \\ &= \underline{r}'_i \end{aligned} \tag{1.75}$$

Non-diagonal elements of **S'** are obtained by the ratios

$$\underline{S}'_{i,j} = \left.\frac{\underline{b}'_i}{\underline{a}'_j}\right|_{\underline{a}'_k=0 \quad \text{for all } k\neq j}, \tag{1.76}$$

Again, their squared magnitudes describe the power transfer from port *j* to port *i*.

Inserting (1.72) and (1.73) into **b'** = **S'**·**a'** yields several helpful conversion formulas. Knowing that (in vector/matrix notation) **V** = **Z**·**I**, it is convenient to have at hand the relationship between a network's impedance matrix **Z** and its **S'** matrix representation. It is easy to show that

$$\mathbf{S'} = \mathbf{D_0} \cdot (\mathbf{Z} - \mathbf{Z_0^*}) \cdot (\mathbf{Z} + \mathbf{Z_0})^{-1} \cdot \mathbf{D_0^{-1}} \tag{1.77}$$

and

$$\mathbf{Z} = \mathbf{D_0^{-1}} \cdot (\mathbf{U} - \mathbf{S'})^{-1} \cdot (\mathbf{Z_0^*} + \mathbf{S'} \cdot \mathbf{Z_0}) \cdot \mathbf{D_0}. \tag{1.78}$$

In a similar manner, from **I** = **Y**·**V**, we get the conversion formulas between the admittance matrix, **Y**, and the generalized scattering matrix, **S'**. Some reshuffling of matrix terms yields

$$\mathbf{S'} = \mathbf{D_0} \cdot (\mathbf{U} - \mathbf{Z_0^*} \cdot \mathbf{Y}) \cdot (\mathbf{U} + \mathbf{Z_0} \cdot \mathbf{Y})^{-1} \cdot \mathbf{D_0^{-1}} \tag{1.79}$$

and

$$\mathbf{Y} = \mathbf{D_0^{-1}} \cdot (\mathbf{Z_0^*} + \mathbf{S'} \cdot \mathbf{Z_0})^{-1} \cdot (\mathbf{U} - \mathbf{S'}) \cdot \mathbf{D_0}, \tag{1.80}$$

or, in terms of the **reference admittance matrix $\mathbf{Y}_0 = \mathbf{Z}_0^{-1}$**,

$$\mathbf{Y} = \mathbf{D_0^{-1}} \cdot \left((\mathbf{Y_0^*})^{-1} + \mathbf{S'} \cdot (\mathbf{Y_0})^{-1}\right)^{-1} \cdot \left(\mathbf{U} - \mathbf{S'}\right) \cdot \mathbf{D_0}. \tag{1.81}$$

Moreover, the question arises how to convert a normalized **S** matrix, measured with matched terminations at all ports (e.g., $Z_{0i} = 50\ \Omega$ for all $i = 1, 2, ..., Q$), into a generalized **S'** matrix with arbitrarily chosen load and source impedances $\underline{\tilde{Z}}_{0i}$. It was shown in [1] that the mismatching effects are to be considered by two $Q \times Q$ diagonal matrices Ξ and Ψ, whose ith diagonal elements are calculated as follows:

$$\underline{\xi}_i = \frac{\underline{\tilde{Z}}_{0i} - \underline{Z}_{0i}}{\underline{\tilde{Z}}_{0i} + \underline{Z}_{0i}} \tag{1.82}$$

and

$$\underline{\psi}_i = \frac{(1 - \underline{\xi}_i^*)\sqrt{1 - |\underline{\xi}_i|^2}}{|1 - \underline{\xi}_i|}. \tag{1.83}$$

For the ith port, $\underline{Z}_{0i}$ is the original normalizing impedance to be converted into the actual one, termed $\underline{\tilde{Z}}_{0i}$. After having available the diagonal elements of Ξ and Ψ, the renormalized **S'** matrix is determined by

$$\mathbf{S'} = \mathbf{\Psi}^{-1} \cdot (\mathbf{S} - \mathbf{\Xi}^*) \cdot (\mathbf{U} - \mathbf{\Xi} \cdot \mathbf{S})^{-1} \cdot \mathbf{\Psi}^*. \tag{1.84}$$

Thus, **S'** matrices allow us to accurately describe the electrical properties of non-matched multi-ports in terms of their individual reflection coefficients and transmission coefficient between their ports.

It is sometimes desirable to check out whether or not a multi-port network is **lossless** and/or **reciprocal**. While reciprocity is easily detected if **S** is a symmetrical matrix (i.e., $\underline{S}_{i,j} = \underline{S}_{j,i}$ for all non-diagonal elements), losslessness requires that the network's scattering matrix be unitary. For a unitary matrix, **S**, we have

$$\mathbf{S}^T \cdot \mathbf{S}^* = \mathbf{U} \tag{1.85}$$

or, equivalently,

$$(\mathbf{S}^T)^{-1} = \mathbf{S}^*. \tag{1.86}$$

In terms of **S** matrix elements, we have to show that each of the sums of all squared magnitudes along any column of **S** is equal to one. In addition, summation of all products of matrix elements in any column by the conjugate of elements in a different column must yield zero. Thus, mathematically, we can restate (1.85) for a Q-port network as

$$\sum_{k=1}^{Q} \underline{S}_{k,i}\underline{S}_{k,j}^* = \delta_{i,j} \quad \text{for all } i,j \le Q \tag{1.87}$$

where

$$\delta_{i,j} = \begin{cases} 1 & \text{if } i = j \\ 0 & \text{if } i \ne j \end{cases} \tag{1.88}$$

designates the **Kronecker delta** symbol.

Next, we investigate the electrical properties of two cascaded multi-port networks with known **S** matrices, $\mathbf{S}^{(1)}$ and $\mathbf{S}^{(2)}$. Suppose the M outputs of the left (N, M) multi-port network, **NET1** in Figure 1-8, are connected to the M inputs of another network, **NET2**, with P outputs. Obviously, the incident scattering parameter vector

$$\mathbf{a}_{\mathbf{M}}^{(2)} = \left(\underline{a}_{N+1}^{(2)}, \underline{a}_{N+2}^{(2)}, \ldots, \underline{a}_{N+M}^{(2)}\right)^{\mathrm{T}} \tag{1.89}$$

of **NET2** is equal to the reflected scattering parameter vector at the output of

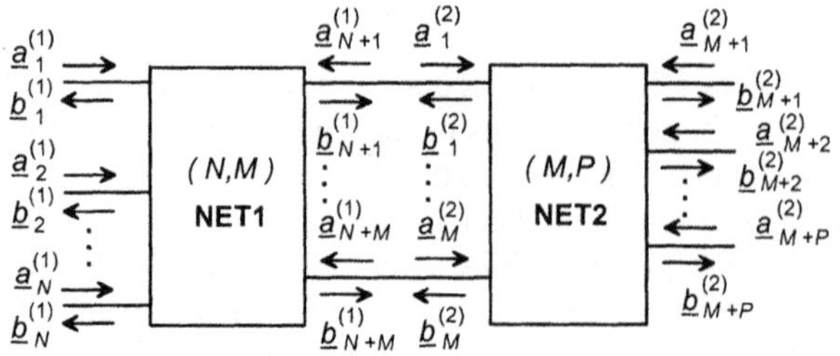

Figure 1-8. Scattering parameters of two cascaded networks.

NET1. Therefore, we have

$$\mathbf{b}_{\mathbf{M}}^{(\mathbf{1})} = \left(\underline{b}_{N+1}^{(1)}, \underline{b}_{N+2}^{(1)}, \ldots, \underline{b}_{N+M}^{(1)}\right)^T = \mathbf{a}_{\mathbf{M}}^{(\mathbf{2})}. \tag{1.90}$$

In a similar manner, we observe that

$$\mathbf{a}_{\mathbf{M}}^{(\mathbf{1})} = \left(\underline{a}_{N+1}^{(1)}, \underline{a}_{N+2}^{(1)}, \ldots, \underline{a}_{N+M}^{(1)}\right)^T \tag{1.91}$$

is equal to

$$\mathbf{b}_{\mathbf{M}}^{(\mathbf{2})} = \left(\underline{b}_{1}^{(2)}, \underline{b}_{2}^{(2)}, \ldots, \underline{b}_{M}^{(2)}\right)^T. \tag{1.92}$$

Note that the total number of ports of **NET1** is $Q^{(1)} = N{+}M$, and that **NET2** has $Q^{(2)} = M{+}P$ ports.

Now, we partition the **S** matrices of **NET1** and **NET2** along the dotted lines as follows:

$$\begin{pmatrix} \underline{b}_1^{(1)} \\ \underline{b}_2^{(1)} \\ \vdots \\ \underline{b}_N^{(1)} \\ \cdots \\ \underline{b}_{N+1}^{(1)} \\ \underline{b}_{N+2}^{(1)} \\ \vdots \\ \underline{b}_{N+M}^{(1)} \end{pmatrix} = \overset{\qquad\qquad N \qquad\qquad\qquad\qquad\qquad M}{\left(\begin{array}{cccc:cccc} \underline{S}_{1,1}^{(1)} & \underline{S}_{1,2}^{(1)} & \cdots & \underline{S}_{1,N}^{(1)} & \underline{S}_{1,N+1}^{(1)} & \underline{S}_{1,N+2}^{(1)} & \cdots & \underline{S}_{1,N+M}^{(1)} \\ \underline{S}_{2,1}^{(1)} & \underline{S}_{2,2}^{(1)} & \cdots & \underline{S}_{2,N}^{(1)} & \underline{S}_{2,N+1}^{(1)} & \underline{S}_{2,N+2}^{(1)} & \cdots & \underline{S}_{2,N+M}^{(1)} \\ \vdots & \vdots & \ddots & \vdots & \vdots & \vdots & \ddots & \vdots \\ \underline{S}_{N,1}^{(1)} & \underline{S}_{N,2}^{(1)} & \cdots & \underline{S}_{N,N}^{(1)} & \underline{S}_{N,N+1}^{(1)} & \underline{S}_{N,N+2}^{(1)} & \cdots & \underline{S}_{N,N+M}^{(1)} \\ \hdashline \underline{S}_{N+1,1}^{(1)} & \underline{S}_{N+1,2}^{(1)} & \cdots & \underline{S}_{N+1,N}^{(1)} & \underline{S}_{N+1,N+1}^{(1)} & \underline{S}_{N+1,N+2}^{(1)} & \cdots & \underline{S}_{N+1,N+M}^{(1)} \\ \underline{S}_{N+2,1}^{(1)} & \underline{S}_{N+2,2}^{(1)} & \cdots & \underline{S}_{N+2,N}^{(1)} & \underline{S}_{N+2,N+1}^{(1)} & \underline{S}_{N+2,N+2}^{(1)} & \cdots & \underline{S}_{N+2,N+M}^{(1)} \\ \vdots & \vdots & \ddots & \vdots & \vdots & \vdots & \ddots & \vdots \\ \underline{S}_{N+M,1}^{(1)} & \underline{S}_{N+M,2}^{(1)} & \cdots & \underline{S}_{N+M,N}^{(1)} & \underline{S}_{N+M,N+1}^{(1)} & \underline{S}_{N+M,N+2}^{(1)} & \cdots & \underline{S}_{N+M,N+M}^{(1)} \end{array}\right)} \cdot \begin{pmatrix} \underline{a}_1^{(1)} \\ \underline{a}_2^{(1)} \\ \vdots \\ \underline{a}_N^{(1)} \\ \cdots \\ \underline{a}_{N+1}^{(1)} \\ \underline{a}_{N+2}^{(1)} \\ \vdots \\ \underline{a}_{N+M}^{(1)} \end{pmatrix} \begin{matrix} \uparrow \\ \\ N \\ \downarrow \\ \cdots \\ \uparrow \\ \\ M \\ \downarrow \end{matrix} \tag{1.93}$$

and

$$\begin{pmatrix} \underline{b}_1^{(2)} \\ \underline{b}_2^{(2)} \\ \vdots \\ \underline{b}_M^{(2)} \\ \cdots \\ \underline{b}_{M+1}^{(2)} \\ \underline{b}_{M+2}^{(2)} \\ \vdots \\ \underline{b}_{M+P}^{(2)} \end{pmatrix} = \begin{pmatrix} \underline{S}_{1,1}^{(2)} & \underline{S}_{1,2}^{(2)} & \cdots & \underline{S}_{1,M}^{(2)} & \vdots & \underline{S}_{1,M+1}^{(2)} & \underline{S}_{1,M+2}^{(2)} & \cdots & \underline{S}_{1,M+P}^{(2)} \\ \underline{S}_{2,1}^{(2)} & \underline{S}_{2,2}^{(2)} & \cdots & \underline{S}_{2,M}^{(2)} & \vdots & \underline{S}_{2,M+1}^{(2)} & \underline{S}_{2,M+2}^{(2)} & \cdots & \underline{S}_{2,M+P}^{(2)} \\ \vdots & \vdots & \ddots & \vdots & \vdots & \vdots & \vdots & \ddots & \vdots \\ \underline{S}_{M,1}^{(2)} & \underline{S}_{M,2}^{(2)} & \cdots & \underline{S}_{M,M}^{(2)} & \vdots & \underline{S}_{M,M+1}^{(2)} & \underline{S}_{M,M+2}^{(2)} & \cdots & \underline{S}_{M,M+P}^{(2)} \\ \cdots & \cdots & \cdots & \cdots & \cdots & \cdots & \cdots & \cdots & \cdots \\ \underline{S}_{M+1,1}^{(2)} & \underline{S}_{M+1,2}^{(2)} & \cdots & \underline{S}_{M+1,M}^{(2)} & \vdots & \underline{S}_{M+1,M+1}^{(2)} & \underline{S}_{M+1,M+2}^{(2)} & \cdots & \underline{S}_{M+1,M+P}^{(2)} \\ \underline{S}_{M+2,1}^{(2)} & \underline{S}_{M+2,2}^{(2)} & \cdots & \underline{S}_{M+2,M}^{(2)} & \vdots & \underline{S}_{M+2,M+1}^{(2)} & \underline{S}_{M+2,M+2}^{(2)} & \cdots & \underline{S}_{M+2,M+P}^{(2)} \\ \vdots & \vdots & \ddots & \vdots & \vdots & \vdots & \vdots & \ddots & \vdots \\ \underline{S}_{M+P,1}^{(2)} & \underline{S}_{M+P,2}^{(2)} & \cdots & \underline{S}_{M+P,M}^{(2)} & \vdots & \underline{S}_{M+P,M+1}^{(2)} & \underline{S}_{M+P,M+2}^{(2)} & \cdots & \underline{S}_{M+P,M+P}^{(2)} \end{pmatrix} \cdot \begin{pmatrix} \underline{a}_1^{(2)} \\ \underline{a}_2^{(2)} \\ \vdots \\ \underline{a}_M^{(2)} \\ \cdots \\ \underline{a}_{M+1}^{(2)} \\ \underline{a}_{M+2}^{(2)} \\ \vdots \\ \underline{a}_{M+P}^{(2)} \end{pmatrix} \tag{1.94}$$

We call the four **sub-matrices** of $\mathbf{S}^{(1)}$, according to their indices in the lower right corners, $\mathbf{S}^{(1)}{}_{\mathbf{N,N}}$, $\mathbf{S}^{(1)}{}_{\mathbf{N,N+M}}$, $\mathbf{S}^{(1)}{}_{\mathbf{N+M,N}}$, and $\mathbf{S}^{(1)}{}_{\mathbf{N+M,N+M}}$. Similarly, those of $\mathbf{S}^{(2)}$ are termed $\mathbf{S}^{(2)}{}_{\mathbf{M,M}}$, $\mathbf{S}^{(2)}{}_{\mathbf{M,M+P}}$, $\mathbf{S}^{(2)}{}_{\mathbf{M+P,M}}$, and $\mathbf{S}^{(2)}{}_{\mathbf{M+P,M+P}}$.

The cascaded network has N inputs and P outputs. Its scattering matrix, $\mathbf{S}^{(C)}$, can also be partitioned into four sub-matrices as follows:

$$\mathbf{S}^{(C)} = \begin{pmatrix} \mathbf{S}^{(C)}_{\mathbf{N,N}} & \vdots & \mathbf{S}^{(C)}_{\mathbf{N,N+P}} \\ \cdots & \cdots & \cdots \\ \mathbf{S}^{(C)}_{\mathbf{N+P,N}} & \vdots & \mathbf{S}^{(C)}_{\mathbf{N+P,N+P}} \end{pmatrix} \tag{1.95}$$

where the four sub-matrices are given by (see, e.g., [1])

$$\mathbf{S}^{(C)}_{\mathbf{N,N}} = \mathbf{S}^{(1)}_{\mathbf{N,N}} + \mathbf{S}^{(1)}_{\mathbf{N,N+M}} \cdot \left(\mathbf{U} - \mathbf{S}^{(2)}_{\mathbf{M,M}} \cdot \mathbf{S}^{(1)}_{\mathbf{N+M,N+M}}\right)^{-1} \cdot \mathbf{S}^{(2)}_{\mathbf{M,M}} \cdot \mathbf{S}^{(1)}_{\mathbf{N+M,N}} \tag{1.96}$$

$$\mathbf{S}^{(C)}_{\mathbf{N,N+P}} = \mathbf{S}^{(1)}_{\mathbf{N,N+M}} \cdot \left(\mathbf{U} - \mathbf{S}^{(2)}_{\mathbf{M,M}} \cdot \mathbf{S}^{(1)}_{\mathbf{N+M,N+M}}\right)^{-1} \cdot \mathbf{S}^{(2)}_{\mathbf{M,M+P}} \tag{1.97}$$

$$\mathbf{S}^{(C)}_{\mathbf{N+P,N}} = \mathbf{S}^{(2)}_{\mathbf{M+P,M}} \cdot \left(\mathbf{U} - \mathbf{S}^{(1)}_{\mathbf{N+M,N+M}} \cdot \mathbf{S}^{(2)}_{\mathbf{M,M}}\right)^{-1} \cdot \mathbf{S}^{(1)}_{\mathbf{N+M,N}} \tag{1.98}$$

$$\mathbf{S}^{(C)}_{\mathbf{N+P,N+P}} = \mathbf{S}^{(2)}_{\mathbf{M+P,M+P}} + \mathbf{S}^{(2)}_{\mathbf{M+P,M}} \cdot \left(\mathbf{U} - \mathbf{S}^{(1)}_{\mathbf{N+M,N+M}} \cdot \mathbf{S}^{(2)}_{\mathbf{M,M}}\right)^{-1} \cdot \mathbf{S}^{(1)}_{\mathbf{N+M,N+M}} \cdot \mathbf{S}^{(2)}_{\mathbf{M,M+P}} \tag{1.99}$$

MATLAB®-based implementations of the above conversion formulas and tests are available on the accompanying CD-ROM. Specifically, the reader will find the following programs and functions ready for practical use:

- **z2s.m** converts a network's impedance matrix, **Z**, into the appertaining generalized **S** matrix. Note that MATLAB function **S = z2s(Z0, Z)** needs as its input arguments not only the elements of a **Z** matrix, but also those of a normalizing reference impedance matrix, $\mathbf{Z}_0$. $\mathbf{Z}_0$ is a diagonal matrix whose elements are either real-valued (e.g., all diagonal elements are $Z_0 = 50\ \Omega$) or complex-valued, i.e., with a capacitive or inductive reactance term.

- **s2z.m** is a conversion program that calculates the **Z** matrix from a given (generalized) **S'** matrix. The function can be called up in a program by adding the following line: **Z = s2z(Z0, S)**.

- **y2s.m** converts a **Y** matrix into the corresponding generalized **S'** matrix. The appropriate function call is **S = y2s(Z0, Y)**.

- **s2y.m** relates a network's generalized **S** matrix to its admittance matrix, **Y**. MATLAB function **Y = s2y(Z0, S)** can be called once $\mathbf{Z}_0$ and the (generalized) **S'** matrix are defined.

- **sn2sg.m** deals with mismatching effects at the Q terminals of a multi-port network. Program **sn2sg.m** needs as its inputs the normalizing impedance matrix $\mathbf{Z}_0$ (denoted Z0, diagonal format!), the mismatched impedance matrix Z0T, and the original normalized **S** matrix, SN. After a few intermediate steps (equations (1.76) - (1.78)), it delivers the generalized **S'** matrix, SG. Hence, the elements of SG take into account the network's actual terminations, including all possible mismatching effects. The MATLAB function call is **SG = sn2sg(Z0, Z0T, SN)**.

- **losstest.m** checks out whether or not a Q-port network is lossless. For that purpose, the network's $Q \times Q$ scattering matrix, **S**, must be specified. Invocation of function **test = losstest(S)** yields a simple verbal statement about the existence or the absence of average power losses occurring in the network.

- **s1s2sc.m** calculates the joint **S** matrix, SC, of two cascaded networks whose scattering matrices are denoted by S1 and S2. Since the first network, **NET1**, has N inputs and M output ports, S1 is a $N \times M$ matrix. The second one, **NET2**, has to have M inputs connected to the M output ports of **NET1**. **NET2** has P output ports. The function is invoked by **SC = s1s2sc(S1, S2)**.

1.4 FUNDAMENTAL PROPERTIES AND MODELS OF MIMO SYSTEMS

As a starting point, we consider a **single-input single-output (SISO) system** with time-dependent input signal, $x(t)$, and output signal, $y(t)$. The system in Figure 1-9 is characterized by an operator, T, that relates output $y(t)$ to the input signal $x(t)$. We define

$$y(t) = \mathrm{T}\{x(t)\}. \tag{1.100}$$

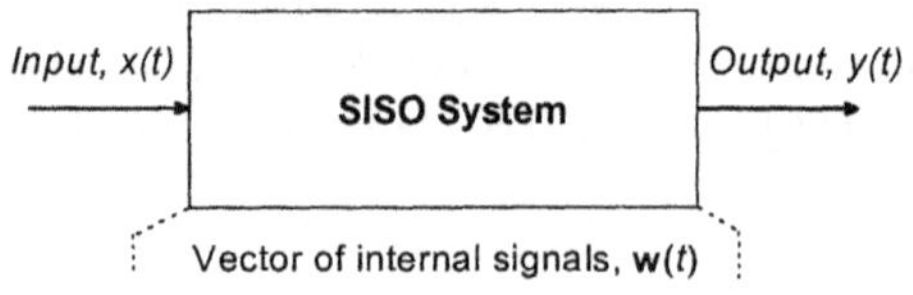

Figure 1-9. Single-input single-output system.

It is straightforward to expand such a SISO system with respect to its number of ports. As we already know, a MIMO system has $N > 1$ inputs and $M > 1$ outputs. Then, analogous to voltages and currents at network ports, we rewrite (1.100) in vector notation. As shown in Figure 1-10, all input signals $x_1(t), x_2(t), \ldots, x_N(t)$ are represented by input column vector **x**(t). Signals $y_1(t)$, $y_2(t), \ldots, y_M(t)$ are jointly denoted by column vector **y**(t). Consequently, the single transform operator T in (1.100) must be replaced by a set of M operators $\mathbf{T} = \{T_1, T_2, \ldots, T_M\}$. The output signals are then given by the following set of equations

$$\begin{aligned} y_1(t) &= T_1\{x_1(t), x_2(t), \cdots x_N(t)\} \\ y_2(t) &= T_2\{x_1(t), x_2(t), \cdots x_N(t)\} \\ &\vdots \\ y_M(t) &= T_M\{x_1(t), x_2(t), \cdots x_N(t)\} \end{aligned} \tag{1.101}$$

Alternatively, using vector notation, we write

$$\mathbf{y}(t) = \mathbf{T}\{\mathbf{x}(t)\}. \tag{1.102}$$

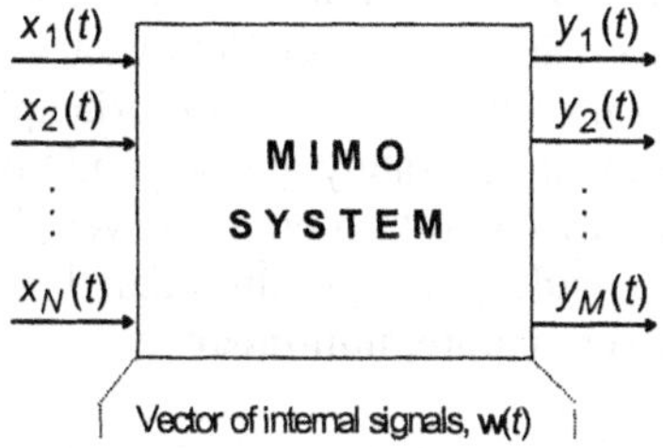

Figure 1-10. Multiple-input multiple-output system.

In system theory it is common to make a few practical assumptions on operators T_i, $i = 1, 2, \ldots M$. Amongst these useful properties are homogeneity, additivity, linearity, time-shift invariance, stability, and causality. Here, we shall extend the standard definitions made for single-input single-output systems, and write them in vector notation for MIMO systems.

Linearity and time invariance play fundamental roles in signal and system analysis, because they allow us to solve, with sufficient accuracy, many technical problems in a fairly straightforward manner. To characterize a given system, we need a few definitions:

Definition 1-1: A MIMO system is **linear** if and only if it is additive and homogeneous. ♦

Definition 1-1a: A MIMO system is **additive** if its output signals, collectively written as a column vector $\mathbf{y}(t) = (y_1(t), y_2(t), \ldots, y_M(t))^T$, due to the sum of two input signal vectors, $\mathbf{x}_\mathbf{A}(t) = (\underline{x}_{A1}(t), \underline{x}_{A2}(t), \ldots, \underline{x}_{AM}(t))^T$ and $\mathbf{x}_\mathbf{A}(t) = (\underline{x}_{A1}(t), \underline{x}_{A2}(t), \ldots, \underline{x}_{AM}(t))^T$ is equal to the sum of $\mathbf{y}_\mathbf{A}(t) = (\underline{y}_{A1}(t), \underline{y}_{A2}(t), \ldots, \underline{y}_{AM}(t))^T$ and $\mathbf{y}_\mathbf{B}(t) = (\underline{y}_{B1}(t), \underline{y}_{B2}(t), \ldots, \underline{y}_{BM}(t))^T$.

Otherwise stated, if inputs $\mathbf{x}_\mathbf{A}(t)$ produce responses $\mathbf{y}_\mathbf{A}(t)$, and the signals of $\mathbf{x}_\mathbf{B}(t)$ generate $\mathbf{y}_\mathbf{B}(t)$, then the summed input signals $\mathbf{x}(t) = \mathbf{x}_\mathbf{A}(t) + \mathbf{x}_\mathbf{B}(t)$ should produce outputs $\mathbf{y}(t) = \mathbf{y}_\mathbf{A}(t) + \mathbf{y}_\mathbf{B}(t)$. ♦

Definition 1-1b: A MIMO system is **homogeneous** if when all inputs, $x(t)$, are scaled by a real or complex constant $\underline{\alpha}$, the response signals, $\mathbf{y}(t)$, will also be scaled by $\underline{\alpha}$. Therefore, if input signal vector $\mathbf{x}_\mathbf{A}(t)$ produces $\mathbf{y}_\mathbf{A}(t)$ then homogeneity requires that the system response to the scaled signals
$\mathbf{x}(t) = \underline{\alpha}\mathbf{x}_\mathbf{A}(t)$ be $\mathbf{y}(t) = \underline{\alpha}\mathbf{y}_\mathbf{A}(t)$. ♦

We can restate Definition 1-1 (including 1-1a and 1-1b) as follows:

A MIMO system is **linear** if and only if it satisfies the **superposition principle**. In short this means if $\mathbf{x}_\mathbf{A}(t) \rightarrow \mathbf{y}_\mathbf{A}(t)$ and $\mathbf{x}_\mathbf{B}(t) \rightarrow \mathbf{y}_\mathbf{B}(t)$, then any linear combination $\mathbf{x}(t) = \underline{\alpha}\mathbf{x}_\mathbf{A}(t) + \underline{\beta}\mathbf{x}_\mathbf{B}(t)$ (with arbitrarily chosen real or complex-valued constants $\underline{\alpha}$ and $\underline{\beta}$) yields output vector $\mathbf{y}(t) = \underline{\alpha}\mathbf{y}_\mathbf{A}(t) + \underline{\beta}\,\mathbf{y}_\mathbf{B}(t)$. An important consequence of the superposition principle is that we can find the outputs due to each input separately and then add or subtract them in an extra step. Therefore, linearity will turn out to be very helpful later on in this text. Note that all other multi-port systems (and there are plenty of them around us!) are, by their very nature, **nonlinear**.

Another fundamental system property is time invariance. It has to be checked using the following statement:

Definition 1-2: A MIMO system is **time invariant** if and only if when $\mathbf{x}_\mathbf{A}(t) \rightarrow \mathbf{y}_\mathbf{A}(t)$ and $\mathbf{x}_\mathbf{B}(t) = \mathbf{x}_\mathbf{A}(t\text{-}t_0) \rightarrow \mathbf{y}_\mathbf{B}(t)$ then, for all time shifts t_0 and arbitrary input vectors $\mathbf{x}_\mathbf{A}(t)$, the system output vector $\mathbf{y}_\mathbf{B}(t)$ is a time-shifted version of $\mathbf{y}_\mathbf{A}(t)$, i.e., $\mathbf{y}_\mathbf{B}(t) = \mathbf{y}_\mathbf{A}(t\text{-}t_0)$. In other words, if the same input signals are presented to the system ports at different times, the resulting output signals will always be the same. ♦

Time-invariant systems are also termed **stationary** or **fixed** because their electrical properties do not change over time. Later on we shall investigate **non-stationary** or **time-variant** systems, too. These systems may or may not be linear. Accordingly, we call them **linear time variant (LTV)** or **nonlinear time-variant (NLTV)** MIMO systems.

Causality is a feature that pertains to physically realizable (i.e., real world) systems only. It has to do with a system's capability, or more precisely its incapability, to anticipate future output values. We can make the following statement about causal multi-port systems:

Definition 1-3: A MIMO system is **causal** if and only if $\mathbf{y}_A(t) = \mathbf{y}_B(t)$ at times $t \le t_0$ given that $\mathbf{x}_A(t) = \mathbf{x}_B(t)$ for $t \le t_0$. Hence, at time t, all internal signals, $\mathbf{w}(t)$, and all outputs, $\mathbf{y}(t)$, only depend upon the system inputs, $\mathbf{x}(t)$, at time t and prior. ♦

Conversely, in a hypothetical **non-causal** system the present outputs, $\mathbf{y}(t)$, would depend on future inputs — something that is not feasible with real electronic circuits.

Another possible classification criterion is a system's capability to store signals. We distinguish between systems with and without memory as follows:

Definition 1-4: A MIMO system is **memoryless**, or **instantaneous**, if and only if all of its outputs, denoted by signal vector $\mathbf{y}(t)$, and all interior signals, $\mathbf{w}(t)$, at any time $t = t_0$ depend on the input signals, $\mathbf{x}(t)$, at the same time $t = t_0$ only. Otherwise, the system is a **dynamic** system. ♦

Definition 1-4a: A **dynamic** MIMO system is termed a **finite-memory** system if and only if its output signals, $\mathbf{y}(t)$, and all internal signals depend upon only those input signals, $\mathbf{x}(t)$, that were applied during the past Δt units of time. ♦

Definition 1-4b: A **dynamic** MIMO system is **non-causal** if at least one of its output signals $\underline{y}_j(t)$ at port j or any one of its internal signals depends on future (i.e., applied at times $t > t_0$) input signals $\underline{x}_i(t)$ at port i. ♦

Examples of memoryless systems include purely resistive networks. Resistors are instantaneous components. Conversely, systems with capacitors and/or inductors have memory. Hence, these systems are easily identified as dynamic systems.

A system can be characterized by the way it accepts and interacts with input signals, and how these input signals are transformed into one or more output signals. There are two main categories, depending on whether the input signals are time-continuous or time-discrete. Thus, we make the following key distinctions:

Definition 1-5: A MIMO system is a **continuous-time** system if and only if all its internal signals, $\mathbf{w}(t)$, and all output signals, $\mathbf{y}(t)$, are continuous-time signals. ♦

Definition 1-6: A MIMO system is a **discrete-time** system if and only if all of its internal state signals, $\mathbf{w}(t)$, and its output signals, $\mathbf{y}(t)$, are discrete-time signals, i.e. all signals involved can change at discrete instants of time $t = kT_S$ only, where T_S is the sampling time interval, and integer numbers $k = ...-2, -1, 0, 1, 2, ...$ are the discrete sampling time indices. ♦

It is often desirable to fully characterize a dynamical system by means of a simple, yet very powerful procedure. For that purpose we employ the so-called unit impulse signal and, consecutively, apply this signal to each one of the MIMO system inputs, while all other inputs are left open. The unit impulse signal, or Dirac delta function, $\delta(t)$, does not conform to the usual mathematical definitions of functions. Mathematically, it belongs to a class of generalized functions, because it is considered as the limit of a conventional function with its width approaching zero (see, e.g., [2]). The total area under $\delta(t)$ is one while its duration approaches zero. Thus, $\delta(t) = 0$ for all $t \neq 0$, and $\delta(0) \rightarrow \infty$. The Dirac delta function has a few very interesting properties that distinguish it from other mathematical functions:

- ***Sifting Property***:

$$\int_{t_1}^{t_2} x(t)\,\delta(t-t_0)\,dt = \begin{cases} x(t_0) & \text{if} \quad t_1 < t_0 < t_2 \\ 0 & \text{otherwise} \end{cases}. \tag{1.103}$$

- ***Sampling Property***: Given a signal $x(t)$, continuous at time t_0, then

$$x(t)\,\delta(t-t_0) = x(t_0)\,\delta(t-t_0). \tag{1.104}$$

- ***Scaling Property***: For arbitrary time scaling factors a and time shifts b, we apply $\delta(at+b) = \frac{1}{|a|}\delta(t+\frac{b}{a})$. (1.105)

Thus, any continuous-time signal $x(t)$ can be described as a continual sum of weighted Dirac impulses. Now let $h(t, \tau)$ denote a linear, continuous-time SISO system's response to such a unit impulse, shifted in time by τ units, i.e., $\delta(-\tau)$. If the system is not only linear but also time-invariant, then $h(t, \tau) = h(t - \tau)$ and, due to the superposition principle, the output signal $y(t)$ can be expressed as a linear combination of all time-shifted impulse responses. We can, therefore, write the output signal in the form of a **convolution integral** as

$$y(t) = \int_{-\infty}^{\infty} x(\tau)h(t-\tau)d\tau . \tag{1.106}$$

In a more concise notation, the convolution integral is commonly written using the convolution symbol * (not to be confused with the complex conjugate!) as follows:

$$y(t) = x(t) * h(t) = h(t) * x(t) . \tag{1.107}$$

Alternatively, we could choose the Laplace transform, $\underline{H}(\underline{s})$, or the Fourier transform, $\underline{H}(j\omega)$, of the system's **impulse response**, $h(t)$. Then, the system input and the output are related by simple products

$$\underline{Y}(\underline{s}) = \underline{H}(\underline{s})\underline{X}(\underline{s}) , \tag{1.108}$$

or

$$\underline{Y}(j\omega) = \underline{H}(j\omega)\underline{X}(j\omega) . \tag{1.109}$$

Having made these preliminary statements about LTI systems with a single input and only one output, we can now extend the concept of convolution to MIMO systems. Hence, we determine an important property of MIMO systems:

Definition 1-7: A linear MIMO system can be unambiguously characterized by a set of impulse responses, $\mathbf{h}(t, \tau)$ in **impulse response matrix** form, measured at the system's output ports. Each element of this matrix is an impulse response signal $h_{j,i}(t, \tau)$, measured at the jth output port with only the ith input port excited by a unit impulse, applied at time τ. ♦

Mathematically, the jth output signal (out of M) of a MIMO system with N inputs and M outputs can be calculated by the sum over N convolution integrals as

$$y_j(t) = \sum_{i=1}^{N} \int_{-\infty}^{\infty} x_i(\tau)h_{j,i}(t,\tau)d\tau \quad j = 1,2,\ldots, M . \tag{1.110}$$

Note that the convolution concept is appropriate for linear time-continuous systems. A more general description for linear, nonlinear, and time-variant MIMO systems can be obtained using **state variables**. As we consider causal systems only, these variables contain information about the past of the system, not about its future. In summary, the state variables of a finite-dimensional system of order P are written in column vector form as

$$\mathbf{w}(t) = \left(w_1(t), w_2(t), \ldots, w_P(t)\right)^T \tag{1.111}$$

Given the MIMO system has N inputs

$$\mathbf{x}(t) = \left(x_1(t), x_2(t), \ldots, x_N(t)\right)^T \tag{1.112}$$

and M outputs

$$\mathbf{y}(t) = \left(y_1(t), y_2(t), \ldots, y_M(t)\right)^T, \tag{1.113}$$

then we can define the **state equations** by a set of simultaneous first-order differential equations where the pth differential equation has the general functional form

$$\dot{w}_p(t) = g\left(w_1(t), w_2(t), \ldots, w_P(t), x_1(t), x_2(t), \ldots, x_N(t), t\right). \tag{1.114}$$

Using vector notation, we write the state equations as

$$\dot{\mathbf{w}}(t) = \mathbf{g}\left(\mathbf{w}(t), \mathbf{x}(t), t\right). \tag{1.115}$$

Similarly, the jth output, $y_j(t)$, of a MIMO system is denoted by

$$y_j(t) = h\left(w_1(t), w_2(t), \ldots, w_P(t), x_1(t), x_2(t), \ldots, x_N(t), t\right). \tag{1.116}$$

Rewriting the system's M output signals in vector form yields a set of **output equations**

$$\mathbf{y}(t) = \mathbf{h}\left(\mathbf{w}(t), \mathbf{x}(t), t\right). \tag{1.117}$$

More specifically, for a **linear time-invariant** MIMO system, (1.114) and (1.116) can be expressed as linear combinations with constant coefficients in the following forms:

$$\dot{w}_P(t) = \sum_{i=1}^{P} a_{p,i}(t) w_i(t) + \sum_{i=1}^{N} b_{p,i}(t) x_i(t), \quad p = 1, 2, \ldots, P, \tag{1.118}$$

$$y_j(t) = \sum_{i=1}^{P} c_{j,i}(t) w_i(t) + \sum_{i=1}^{N} d_{j,i}(t) x_i(t), \quad j = 1, 2, \ldots, M. \tag{1.119}$$

Using vectors and matrices, the **state equations** and the **output equations** are obtained as

$$\dot{\mathbf{w}}(t) = \mathbf{A}(t) \cdot \mathbf{w}(t) + \mathbf{B}(t) \cdot \mathbf{x}(t), \tag{1.120}$$

$$\mathbf{y}(t) = \mathbf{C}(t) \cdot \mathbf{w}(t) + \mathbf{D}(t) \cdot \mathbf{x}(t). \tag{1.121}$$

Since we assume the system has N inputs, M outputs, and P state variables, it should be clear that $\dot{\mathbf{w}}(t)$ and $\mathbf{w}(t)$ are both $P \times 1$ column vectors, whereas the formats of vectors $\mathbf{x}(t)$ and $\mathbf{y}(t)$ are $N \times 1$ and $M \times 1$, respectively. Therefore, we find that

- **A** is a $P \times P$ square state coupling matrix,
- **B** is a $P \times N$ matrix,
- **C** is a $M \times P$ matrix, and
- **D** is a $M \times N$ matrix.

Two-port networks (SISO systems) represent a special case where, depending on the number of state variables, **A** is the only matrix (size $P \times P$). **B** and **C** are row and column vectors, respectively, and **D** reduces to a scalar quantity. All matrix or vector elements may, in principle, vary as functions of any sets of free parameters. As we shall see, it is sufficient for most of our investigations to narrow the degrees of freedom to time, *t*, as this is the only free parameter. Hence, in a **time-varying** or **non-autonomous system** the matrix elements change over time whereas in an LTI system they are constant.

Solving the set of first-order differential equations (1.120) for the vector of state variables, $\mathbf{w}(t)$, yields the sum of the initial condition (or zero-input) solution plus the forced (or zero-state) response. Thus, the solution of the state equation is given by (see, e.g., [3])

$$\mathbf{w}(t) = \mathbf{exp}(\mathbf{A}t) \cdot \mathbf{w}(0^-) + \int_{0^-}^{t} \mathbf{exp}(\mathbf{A}(t-\tau)) \cdot \mathbf{B} \cdot \mathbf{x}(\tau) d\tau \tag{1.122}$$

where column vector $\mathbf{w}(0^-)$ represents the initial state conditions at time 0^-. The second term in the sum on the right of (1.122) can be considered as a convolution integral of matrix $\mathbf{B} \cdot \mathbf{x}(t)$ and a $P \times P$ matrix exponential, termed **state transition matrix**,

$$\mathbf{\Phi}(t) = \mathbf{exp}(\mathbf{A}t) = \sum_{i=0}^{\infty} \frac{\mathbf{A}^i t^i}{i!} = \mathbf{U} + \mathbf{A}t + \frac{\mathbf{A}^2 t^2}{2!} + \ldots \ . \tag{1.123}$$

For our further investigations it is important to realize (see, e.g., [4]) that the Laplace transform, denoted by operator L{...}, of the state transition matrix is obtained in the form of a matrix inverse as

$$\mathbf{\Phi}(\underline{s}) = \mathrm{L}\{\mathbf{\Phi}(t)\} = (\underline{s}\mathbf{U} - \mathbf{A})^{-1} . \tag{1.124}$$

Similarly, solving the set of M **output equations** (1.121) for vector $\mathbf{y}(t)$ yields

$$\mathbf{y}(t) = \mathbf{C} \cdot \mathbf{\Phi}(t) \cdot \mathbf{w}(0^-) + \int_{0^-}^{t} \mathbf{C} \cdot \mathbf{\Phi}(t-\tau) \cdot \mathbf{B} \cdot \mathbf{x}(\tau) d\tau + \mathbf{D} \cdot \mathbf{x}(t) \tag{1.125}$$

where, again, the integral term in the sum is a convolution of $\mathbf{B} \cdot \mathbf{x}(t)$ with state transition matrix $\mathbf{\Phi}(t)$, pre-multiplied by the system's $M \times P$ matrix, $\mathbf{C}$.

Equation (1.125) can be rewritten as

$$\mathbf{y}(t) = \mathbf{C} \cdot \left[\mathbf{\Phi}(t) \cdot \mathbf{w}(0^-) + \int_{0^-}^{t} \mathbf{\Phi}(t-\tau) \cdot \mathbf{B} \cdot \mathbf{x}(\tau) d\tau \right] + \mathbf{D} \cdot \mathbf{x}(t) . \tag{1.126}$$

It is sometimes preferable to express the state transition matrix $\mathbf{\Phi}(t)$ = $\mathbf{exp}(\mathbf{A}t)$ in the form of a finite sum instead of writing it as an infinite sum as in (1.123). Given an arbitrary $P \times P$ square matrix $\mathbf{A}$, we can use the Cayley-Hamilton theorem [see, e.g., Appendix C of [2]) to calculate any function, $f(\mathbf{A})$, of that matrix as a linear combination

$$f(\mathbf{A}) = \sum_{p=1}^{P-1} \underline{\gamma}_p \mathbf{A}^p \tag{1.127}$$

where P-1 coefficients $\underline{\gamma}_p$ are to be calculated.

The **Cayley-Hamilton theorem** states that any matrix satisfies its own characteristic equation. That is,

$$f(\lambda) = \det(\mathbf{A} - \lambda\mathbf{U}) = 0 \quad \Leftrightarrow \quad f(\mathbf{A}) = \mathbf{0} \tag{1.128}$$

for P real or complex-valued **eigenvalues,** $\underline{\lambda}_p$, of matrix **A**.

If, from $f(\lambda) = \det(\mathbf{A} - \lambda\mathbf{U}) = 0$, we know the P distinct eigenvalues of matrix **A**, we determine $p = 0, 1, ..., P$-1 unknown coefficients, $\underline{\gamma}_p$, by the following set of equations

$$f(\underline{\lambda}_1) = \underline{\gamma}_0 + \underline{\gamma}_1 \underline{\lambda}_1 + \underline{\gamma}_2 \underline{\lambda}_1^2 + \cdots + \underline{\gamma}_{P-1} \underline{\lambda}_1^{P-1}$$

$$f(\underline{\lambda}_2) = \underline{\gamma}_0 + \underline{\gamma}_1 \underline{\lambda}_2 + \underline{\gamma}_2 \underline{\lambda}_2^2 + \cdots + \underline{\gamma}_{P-1} \underline{\lambda}_2^{P-1}$$

$$\vdots$$

$$f(\underline{\lambda}_P) = \underline{\gamma}_0 + \underline{\gamma}_1 \underline{\lambda}_P + \underline{\gamma}_2 \underline{\lambda}_P^2 + \cdots + \underline{\gamma}_{P-1} \underline{\lambda}_P^{P-1}. \tag{1.129}$$

We observe that for any state transition matrix, $\mathbf{\Phi}(t)$, the terms on the left of (1.129) are time-dependent scalars $\exp(\underline{\lambda}_p t)$ where the $\underline{\lambda}_p$'s are pre-calculated eigenvalues of **A**. Once we have solved (1.129) for the time-dependent coefficients $\underline{\gamma}_p(t)$, they can be entered into the finite sum of (1.127). If the P eigenvalues of **A** are *not* distinct, the equation corresponding to the repeated eigenvalue has to be differentiated with respect to λ.

Finally, it is more convenient to calculate simple products rather than convolution integrals. This is the reason why we usually prefer to take Laplace transforms on both sides of (1.122) and (1.126). Hence, we obtain the Laplace-transformed version of the **state equations** as

$$\mathbf{W}(\underline{s}) = (\underline{s}\mathbf{U} - \mathbf{A})^{-1} \cdot \mathbf{w}(0^-) + (\underline{s}\mathbf{U} - \mathbf{A})^{-1} \cdot \mathbf{B} \cdot \mathbf{X}(\underline{s}). \tag{1.130}$$

Note that the first term of the sum, called zero-input component of the state vector, vanishes if the initial state variables are zero, that is $\mathbf{w}(0^-) = \mathbf{0}$.

By taking Laplace transforms on both sides of the **output equations** in time domain (1.126), we obtain

$$\mathbf{Y}(\underline{s}) = \mathbf{C} \cdot (\underline{s}\mathbf{U} - \mathbf{A})^{-1} \cdot \mathbf{w}(0^-) + (\mathbf{C} \cdot (\underline{s}\mathbf{U} - \mathbf{A})^{-1} \cdot \mathbf{B} + \mathbf{D}) \cdot \mathbf{X}(\underline{s}). \tag{1.131}$$

If we set all initial state variables $\mathbf{w}(0^-)$ equal to zero then the Laplace-transformed output vector, $\mathbf{Y}(\underline{s})$, reduces to

$$\mathbf{Y}(\underline{s}) = (\mathbf{C} \cdot (\underline{s}\mathbf{U} - \mathbf{A})^{-1} \cdot \mathbf{B} + \mathbf{D}) \cdot \mathbf{X}(\underline{s}). \tag{1.132}$$

We rewrite this as

$$\mathbf{Y}(\underline{s}) = \mathbf{H}(\underline{s}) \cdot \mathbf{X}(\underline{s}), \tag{1.133}$$

and call

$$\mathbf{H}(\underline{s}) = \mathbf{C} \cdot (\underline{s}\mathbf{U} - \mathbf{A})^{-1} \cdot \mathbf{B} + \mathbf{D} \tag{1.134}$$

the MIMO system's $M \times N$ **transfer matrix**. Its elements are **transfer functions**, $\underline{H}_{j,i}(s)$, relating Laplace transforms of the jth output to the ith input by the ratio

$$\underline{H}_{j,i}(\underline{s}) = \frac{\underline{Y}_j(\underline{s})}{\underline{X}_i(\underline{s})}. \tag{1.135}$$

It should be mentioned that each $\underline{H}_{j,i}(s)$ assumes zero initial conditions and that all inputs, except for the ith one, are zero. Referring back to Definition 1-7, we note that $\underline{H}_{j,i}(s)$ is the Laplace-transformed representation of the impulse response $h_{j,i}(t)$. Conversely, we can relate the **impulse response matrix** to the **transfer matrix** by means of the inverse Laplace transform as follows:

$$\mathbf{h}(t) = \mathcal{L}^{-1}\{\mathbf{H}(\underline{s})\}. \tag{1.136}$$

In the general case, $\mathbf{H}(\underline{s})$ is a three-dimensional transfer matrix with M rows and P columns. Its depth is equal to the number of $\underline{s}$-polynomial coefficients of the MIMO system.

EXAMPLE 1-5: Consider the electrical circuit shown in Figure 1-11. It has two input voltages, $v_1(t)$ and $v_2(t)$, and a single output voltage, $v_3(t)$. Counting the number of energy-storing elements (i.e., capacitor C and inductor L) yields $P = 2$ state variables, denoted $w_1(t)$ and $w_2(t)$. We assume zero initial conditions and identify the input variables as $x_1(t) = v_1(t)$ and $x_2(t) = v_2(t)$, respectively. By definition, we consider $y(t) = v_3(t)$ as the output variable.

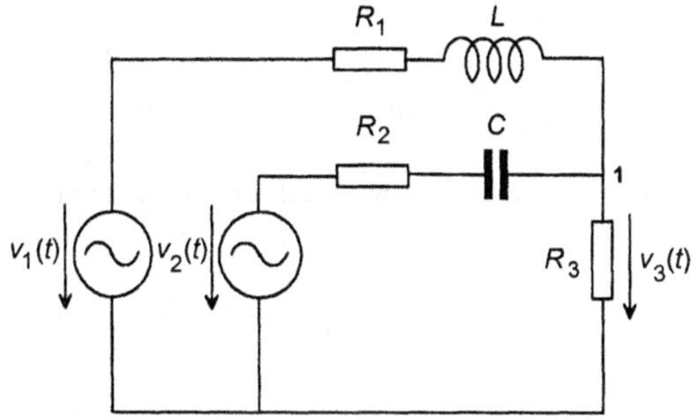

Figure 1-11. Two-input single-output system of example 1-5.

We know that the voltage across inductor L, $v_L(t)$, is related to current $i_L(t)$ and to the first state variable, $w_1(t)$, by

$$v_L(t) = L\frac{di_L(t)}{dt} = L\dot{w}_1(t)\,. \tag{1.137}$$

The current $i_C(t)$ flowing through capacitor C is related to voltage $v_C(t)$ that serves as the second state variable, $w_2(t)$, by

$$i_L(t) = C\frac{dv_C(t)}{dt} = C\dot{w}_2(t)\,. \tag{1.138}$$

Next, we apply Kirchhoff's current law with respect to node 1. Current $i_{R3}(t)$ through resistor R_3 is obtained by summation of currents $i_L(t)$ and $i_C(t)$. Therefore, the output voltage is given by

$$y(t) = v_3(t) = R_3\, i_{R3}(t) = R_3\,(w_1(t) + C\,\dot{w}_2(t))\,. \tag{1.139}$$

By virtue of Kirchhoff's voltage law, the algebraic sum of all voltages across elements in the two loops must be zero. Therefore, in the outer loop, we have

$$\begin{aligned} & v_1(t) - v_3(t) - v_L(t) - v_{R1}(t) = \\ & x_1(t) - (R_3\, w_1(t) + R_3\, C\, \dot{w}_2(t)) - L\, \dot{w}_1(t) - R_1\, w_1(t) = 0, \end{aligned} \tag{1.140}$$

and, similarly, the sum of all voltages in the inner loop is

$$\begin{aligned} & v_2(t) - v_3(t) - v_C(t) - v_{R2}(t) = \\ & x_2(t) - (R_3\, w_1(t) + R_3\, C\, \dot{w}_2(t)) - w_2(t) - R_2\, C\, \dot{w}_2(t) = 0. \end{aligned} \tag{1.141}$$

Solving (1.141) for $\dot{w}_2(t)$ yields the second one of the two state equations. We obtain

$$\dot{w}_2(t) = -\frac{R_3}{(R_2 + R_3)C} w_1(t) - \frac{1}{(R_2 + R_3)C} w_2(t) + 0x_1(t) + \frac{1}{(R_2 + R_3)C} x_2(t)\,. \tag{1.142}$$

Substituting $\dot{w}_2(t)$ into (1.139) and (1.140), we find that the first state equation is given by

$$\dot{w}_1(t) = \left(\frac{R_3^2}{R_2 + R_3} - R_1 - R_3\right)\frac{1}{L} w_1(t) + \frac{R_3}{(R_2 + R_3)L} w_2(t) + \frac{1}{L} x_1(t) - \frac{R_3}{(R_2 + R_3)L} x_2(t)\,, \tag{1.143}$$

and the output equation is obtained as

$$y(t) = R_3\left(1 - \frac{R_3}{R_2 + R_3}\right) w_1(t) - \frac{R_3}{R_2 + R_3} w_2(t) + 0x_1(t) + \frac{R_3}{R_2 + R_3} x_2(t)\,. \tag{1.144}$$

Recalling equations (1.120) and (1.121), we note that for the circuit under consideration all elements of matrices **A**, **B**, **C**, and **D** are constant over time. Thus, the state-space model for our circuit can be written in vector/matrix form as

$$\dot{\mathbf{w}}(t) = \mathbf{A} \cdot \mathbf{w}(t) + \mathbf{B} \cdot \mathbf{x}(t)\,, \tag{1.145}$$

jointly with the output equation (note the **scalar** output variable, $y(t)$!)

$$y(t) = \mathbf{C} \cdot \mathbf{w}(t) + \mathbf{D} \cdot \mathbf{x}(t) \tag{1.146}$$

where the 2×2 state matrices, **A** and **B**, and 1×2 row vectors, **C** and **D,** are

$$\mathbf{A} = \begin{pmatrix} \left(\dfrac{R_3^2}{R_2+R_3} - R_1 - R_3\right)\dfrac{1}{L} & \dfrac{R_3}{(R_2+R_3)L} \\ -\dfrac{R_3}{(R_2+R_3)C} & -\dfrac{1}{(R_2+R_3)C} \end{pmatrix}, \tag{1.147}$$

$$\mathbf{B} = \begin{pmatrix} \dfrac{1}{L} & -\dfrac{R_3}{(R_2+R_3)L} \\ 0 & \dfrac{1}{(R_2+R_3)C} \end{pmatrix}, \tag{1.148}$$

$$\mathbf{C} = \begin{pmatrix} R_3\left(1 - \dfrac{R_3}{R_2+R_3}\right) & -\dfrac{R_3}{R_2+R_3} \end{pmatrix}, \tag{1.149}$$

$$\mathbf{D} = \begin{pmatrix} 0 & \dfrac{R_3}{R_2+R_3} \end{pmatrix}. \tag{1.150}$$

Substituting these expressions into (1.134), we find the 1×2 row vector $\mathbf{H}(\underline{s})$ composed of two transfer functions, $\underline{H}_{3,1}(\underline{s})$ and $\underline{H}_{3,2}(\underline{s})$, as

$$\mathbf{H}(\underline{s}) = \mathbf{C}\cdot(\underline{s}\mathbf{U}-\mathbf{A})^{-1}\cdot\mathbf{B}+\mathbf{D} = \begin{bmatrix} \underline{H}_{3,1}(\underline{s}) & \underline{H}_{3,2}(\underline{s}) \end{bmatrix} = \begin{bmatrix} \dfrac{\underline{Y}_3(\underline{s})}{\underline{X}_1(\underline{s})} & \dfrac{\underline{Y}_3(\underline{s})}{\underline{X}_2(\underline{s})} \end{bmatrix}. \tag{1.151}$$

Note that the Laplace transform of the output $\underline{Y}_3(\underline{s})$ equals $\underline{Y}(\underline{s})$. In terms of the circuit parameters we find

$$\underline{H}_{3,1}(\underline{s}) = \frac{1+\underline{s}R_2C}{\underline{s}^2LC(1+\frac{R_2}{R_3})+\underline{s}(C(R_1+R_2+\frac{R_1R_2}{R_3})+\frac{L}{R_3})+(1+\frac{R_1}{R_3})} \tag{1.152}$$

and

$$\underline{H}_{3,2}(\underline{s}) = \frac{\underline{s}^2LC+\underline{s}R_1C}{\underline{s}^2LC(1+\frac{R_2}{R_3})+\underline{s}(C(R_1+R_2+\frac{R_1R_2}{R_3})+\frac{L}{R_3})+(1+\frac{R_1}{R_3})}. \tag{1.153}$$

▶◀

For **discrete-time** MIMO systems, the **state equations** are written using discrete times, kT_S or time increment k for short, instead of continuous-time variable, t. Therefore, the set of first-order *differential* equations (1.120) turns into a set of first-order *difference* equations, and we obtain

$$\dot{\mathbf{w}}(k+1) = \mathbf{A}(k) \cdot \mathbf{w}(k) + \mathbf{B}(k) \cdot \mathbf{x}(k) . \tag{1.154}$$

Similarly, the **output equations** of (1.121) are substituted by their discrete-time equivalent

$$\mathbf{y}(k) = \mathbf{C}(k) \cdot \mathbf{w}(k) + \mathbf{D}(k) \cdot \mathbf{x}(k) . \tag{1.155}$$

By a simple interchange of complex Laplace variable $\underline{s}$ with transform variable

$$\underline{z} = exp(\underline{s}\, T_S)\Big|_{\underline{s} = j\omega} , \tag{1.156}$$

we obtain the equivalent frequency or $\underline{z}$-domain representation of the state equations as

$$\mathbf{W}(\underline{z}) = (\underline{z}\mathbf{U} - \mathbf{A})^{-1} \cdot \mathbf{w}(0) + (\underline{z}\mathbf{U} - \mathbf{A})^{-1} \cdot \mathbf{B} \cdot \mathbf{X}(\underline{z}) , \tag{1.157}$$

and the $\underline{z}$-domain description of the output vector is given by

$$\mathbf{Y}(\underline{z}) = \mathbf{C} \cdot (\underline{z}\mathbf{U} - \mathbf{A})^{-1} \cdot \mathbf{w}(0) + (\mathbf{C} \cdot (\underline{z}\mathbf{U} - \mathbf{A})^{-1} \cdot \mathbf{B} + \mathbf{D}) \cdot \mathbf{X}(\underline{z}) . \tag{1.158}$$

Then, again assuming zero initial conditions, we get the **transfer matrix** as

$$\mathbf{H}(\underline{z}) = \mathbf{C} \cdot (\underline{z}\mathbf{U} - \mathbf{A})^{-1} \cdot \mathbf{B} + \mathbf{D}. \tag{1.159}$$

Its elements are $\underline{z}$-domain **transfer functions** relating, under zero initial conditions, the $\underline{z}$-transform of the jth response (= system output) to the $\underline{z}$-transform of the ith excitation (= system input) by the ratio

$$\underline{H}_{j,i}(\underline{z}) = \frac{\underline{Y}_j(\underline{z})}{\underline{X}_i(\underline{z})} . \tag{1.160}$$

A discrete-time and causal LTI-type **SISO** system with zero initial conditions would then be described by the difference equation

$$\sum_{n=0}^{N} a_n y(k-n) = \sum_{m=0}^{M} b_m x(k-m) \tag{1.161}$$

where $x(k)$ is the sequence of input samples and $y(k)$ is the output sample sequence. Coefficients $a_0, a_1, \ldots, a_N$ and $b_0, b_1, \ldots, b_M$ are two distinct sets of real-valued numbers. If the present output should not depend on future inputs we need causality which means that $M \le N$.

By taking $\underline{z}$-transforms on both sides of (1.161), we can write the SISO system's transfer function as the ratio

$$\underline{H}(\underline{z}) = \frac{\underline{Y}(\underline{z})}{\underline{X}(\underline{z})} = \frac{\sum_{m=0}^{M} b_m \underline{z}^{-m}}{\sum_{n=0}^{N} a_n \underline{z}^{-n}} = \frac{b_0 + b_1\underline{z}^{-1} + b_2\underline{z}^{-2} + \cdots + b_M\underline{z}^{-M}}{a_0 + a_1\underline{z}^{-1} + a_2\underline{z}^{-2} + \cdots + a_N\underline{z}^{-N}}, \quad M \le N. \tag{1.162}$$

By dividing the numerator and the denominator polynomial by a_0, we shall generally set a_0=1. It may be useful to convert this system representation into the state-space model. To do this, we write a discrete-time version of the SISO system's state equations in the form of

$$\begin{aligned}\mathbf{w}(k+1) &= \begin{pmatrix} 0 & 1 & 0 & \cdots & 0 \\ 0 & 0 & 1 & \cdots & 0 \\ \vdots & \vdots & \vdots & \vdots & \vdots \\ 0 & 0 & 0 & \cdots & 1 \\ -a_N & -a_{N-1} & \cdots & -a_2 & -a_1 \end{pmatrix} \mathbf{w}(k) + \begin{pmatrix} 0 \\ 0 \\ \vdots \\ 0 \\ 1 \end{pmatrix} x(k) \\ &= \mathbf{A} \cdot \mathbf{w}(k) + \mathbf{B}\, x(k). \end{aligned} \tag{1.163}$$

Note that $\mathbf{w}(k)$ and $\mathbf{w}(k+1)$ are $P\times 1$ column vectors whereas $x(k)$ represents a scalar sequence of input samples. $\mathbf{A}$ is a square $N\times N$ matrix, and $\mathbf{B}$ takes on the form of a $N\times 1$ column vector whose elements are all zero except for the last one. It should be noted that N, the degree of the denominator polynomial of $\underline{H}(\underline{z})$, is identical to the number P of system state variables.

Similarly, if we make $M = N$, the output equation is given as a function of the $\underline{z}$-polynomial coefficients by

$$y(k) = (b_M - b_0 a_N, b_{M-1} - b_0 a_{N-1}, \cdots$$
$$\cdots, b_2 - b_0 a_2, b_1 - b_0 a_1) \cdot \mathbf{w}(k) + b_0 x(k) \quad (1.164)$$
$$= \mathbf{C} \cdot \mathbf{w}(k) + \mathbf{D} x(k).$$

In view of (1.155), we notice that $\mathbf{C}$ is a $1 \times N$ row vector and $\mathbf{D} = b_0$ is a scalar.

With the aid of the above results it is easy to derive the state-space model of a continuous-time SISO system from its transfer function

$$\underline{H}(\underline{s}) = \frac{\underline{Y}(\underline{s})}{\underline{X}(\underline{s})} = \frac{\sum_{m=0}^{M} b_m \underline{s}^m}{\sum_{n=0}^{N} a_n \underline{s}^n} = \frac{b_0 + b_1 \underline{s} + b_2 \underline{s}^2 + \cdots + b_M \underline{s}^M}{a_0 + a_1 \underline{s} + a_2 \underline{s}^2 + \cdots + a_N \underline{s}^N}, \quad M \leq N. \quad (1.165)$$

Accordingly, the state equations are given by

$$\dot{\mathbf{w}}(t) = \begin{pmatrix} 0 & 1 & 0 & \cdots & 0 \\ 0 & 0 & 1 & \cdots & 0 \\ \vdots & \vdots & \vdots & \vdots & \vdots \\ 0 & 0 & 0 & \cdots & 1 \\ -a_0 & -a_1 & -a_2 & \cdots & -a_{N-1} \end{pmatrix} \mathbf{w}(t) + \begin{pmatrix} 0 \\ 0 \\ \vdots \\ 0 \\ 1 \end{pmatrix} x(t) \quad (1.166)$$
$$= \mathbf{A} \cdot \mathbf{w}(t) + \mathbf{B}\, x(t),$$

and, provided that $N = M$, the output equation can be written as

$$y(t) = (b_0 - a_0 b_N, b_1 - a_1 b_N, \cdots$$
$$\cdots, b_{N-2} - a_{N-2} b_N, b_{N-1} - a_{N-1} b_N) \cdot \mathbf{w}(t) + b_N x(t) \quad (1.167)$$
$$= \mathbf{C} \cdot \mathbf{w}(t) + \mathbf{D} x(t).$$

Now let us suppose that the system has a single input and multiple outputs. The following example shall explain the conversion method. Of course, similar analyses can be performed if we need the state-space model of a continuous-time SIMO system.

EXAMPLE 1-6: Consider a discrete time LTI system with one input and two outputs. We assume zero initial conditions. The SIMO system's transfer function matrix is chosen as follows

$$\mathbf{H}(\underline{z}) = \frac{1}{\underline{X}(\underline{z})}\begin{pmatrix} \underline{Y}_1(\underline{z}) \\ \underline{Y}_2(\underline{z}) \end{pmatrix} = \frac{1}{1+2\underline{z}^{-1}-3\underline{z}^{-2}}\begin{pmatrix} 3-4\underline{z}^{-1} \\ 2+3\underline{z}^{-1}+\underline{z}^{-2} \end{pmatrix}. \tag{1.168}$$

Since the coefficients of the common denominator polynomial are $a_0=1$, $a_1=2$, and $a_2 = -3$, application of (1.163) yields the state equations

$$\begin{aligned} \mathbf{w}(k+1) &= \begin{pmatrix} 0 & 1 \\ 3 & -2 \end{pmatrix}\mathbf{w}(k) + \begin{pmatrix} 0 \\ 1 \end{pmatrix}x(k) \\ &= \mathbf{A}\cdot\mathbf{w}(k) + \mathbf{B}\,x(k). \end{aligned} \tag{1.169}$$

In the special case of the numerator being $\underline{Y}_1(\underline{z})$ we have $b_{0,1} = 3$, $b_{1,1} = -4$, and $b_{2,1} = 0$. For $\underline{H}_2(\underline{z})$, the numerator coefficients are $b_{0,2} = 2$, $b_{1,2} = 3$, and $b_{2,2} = 1$. Substituting these two sets of coefficients into (1.164), we find that the output equations are

$$\begin{aligned} \mathbf{y}(k) = \begin{pmatrix} y_1(k) \\ y_2(k) \end{pmatrix} &= \begin{pmatrix} 9 & -10 \\ 7 & -1 \end{pmatrix}\cdot\begin{pmatrix} w_1(k) \\ w_2(k) \end{pmatrix} + \begin{pmatrix} 3 \\ 2 \end{pmatrix}x(k) \\ &= \mathbf{C}\cdot\mathbf{w}(k) + \mathbf{D}\,x(k). \end{aligned} \tag{1.170}$$

►◄

Most of the problems associated with conversions between linear system models are addressed by MATLAB toolboxes. Specifically, in the context of this chapter, the Signal Processing Toolbox [5] and the Control System Toolbox [6] offer several powerful mathematical tools. Among them are MATLAB programs **"tf2ss.m"** and **"ss2tf.m"** for conversion between transfer functions and state-space matrices, and vice-versa. Program **"c2d.m"** performs conversion of state-space models from continuous to discrete time assuming a zero-order hold on the inputs and specified sampling time T_S. **"d2c.m"** does the reverse.

1.5 CHAPTER 1 PROBLEMS

1.1 Consider the four-port network shown in Figure 1-4 with all resistances $R = 10\ \Omega$ and admittances $1/\omega C = 20\ \Omega$. Assume that port 2 is the input and port 4 the output of a two-port network.

(a) Calculate the reduced 2×2 matrix, $\widetilde{\mathbf{Y}}$, of the two-port circuit.

(b) Determine the two-port's voltage transfer function, $\underline{H}_V$.

(c) Use MATLAB program "**mimo2t.m**" to calculate the components of an equivalent two-port T-network.

1.2 A purely resistive two-port T-network, as shown in Figure 1-3, shall be used as a matched 10-dB attenuator (log. power ratio!) with a 50-Ω characteristic impedance.

(a) Determine the set of optimum resistances $\underline{Z}_{T1} = R_{T1}$, $\underline{Z}_{T2} = R_{T2}$, and $\underline{Z}_{T3} = R_{T3}$ that yield minimum reflection at both ports.

(b) How can this T-network be converted into an equivalent two-port π-network?

(c) Find the set of optimum components for a 75-Ω characteristic impedance.

1.3 Consider a four-port network with normalized scattering matrix

$$\mathbf{S} = \begin{pmatrix} 0.2\angle 30^\circ & 0.3\angle 45^\circ & 0 & 0.8\angle 45^\circ \\ 0.3\angle 45^\circ & 0.7\angle -45^\circ & 0.9\angle 30^\circ & 0 \\ 0 & 0.9\angle 30^\circ & 0 & 0.2\angle -10^\circ \\ 0.8\angle 45^\circ & 0 & 0.2\angle -10^\circ & 0.2\angle -90^\circ \end{pmatrix}.$$

(a) Determine whether or not this network is reciprocal.

(b) Use MATLAB program **"losstest.m"** to test if the network is lossless.

(c) Calculate the reflection coefficient at port 2 if port 4 is short-circuited and all other ports are matched.

1.4 We consider a cascade of two multi-port networks, **NET1** and **NET2**, with the two **S** matrices $\mathbf{S}^{(1)} = \begin{pmatrix} 0.3 & 0.1 & 0.2 & 0.4 \\ 0.1 & 0.5 & 0.2 & 0.3 \\ 0.2 & 0.2 & 0.9 & 0.1 \\ 0.4 & 0.3 & 0.1 & 0.7 \end{pmatrix}$ and $\mathbf{S}^{(2)} = \begin{pmatrix} 0.2\angle 0 & 0.7\angle 90^\circ \\ 0.7\angle 90^\circ & 0.3\angle 0 \end{pmatrix}$.

Superscripts (1) and (2) denote **NET1** and **NET2**, respectively. **NET2** is connected to ports 3 and 4 of **NET1**.

(a) Write down the two sets of equations (linear combinations) that relate scattering variables $\mathbf{a}^{(1)}$ to $\mathbf{b}^{(1)}$ for **NET1**, and those relating the elements of vector $\mathbf{a}^{(2)}$ to $\mathbf{b}^{(2)}$ of **NET2**.

(b) Use MATLAB program "**s2y.m**" to calculate admittance matrices $\mathbf{Y}^{(1)}$ and $\mathbf{Y}^{(2)}$ of **NET1** and **NET2**.

(c) Partition scattering matrix $\mathbf{S}^{(1)}$ into four appropriate sub-matrices, and calculate, using MATLAB program **"s1s2sc.m"**, the scattering matrix, $\mathbf{S}^{(C)}$, of the cascaded network.

1.5 Determine whether or not an ideal mixing device, modeled as a mathematically ideal multiplier, is

(a) a linear, and/or

(b) a time-invariant device.

1.6 Consider the four-port network shown in Figure 1-4 with resistance R = 100 kΩ and capacitance C = 100 nF. Assume that port 1 is the input and port 3 the output of a two-port network. Take the node-to-ground voltages as the state variables, and input current $\underline{I}_1(t)$ at port 1 as the first output variable. Let output voltage $\underline{V}_3(t)$ at port 3 be the second output variable.

(a) Find the state-space description of the two-port network.
(b) Solve the state equations for state vector $\mathbf{w}(t)$, assuming zero initial state conditions.
(c) Solve the output equations for current $\underline{I}_1(t)$ and voltage $\underline{V}_3(t)$.
(d) Calculate the state transition matrix $\mathbf{\Phi}(t)$ of the given two-port network.

1.7 A continuous-time SIMO system has one input and two outputs. Its transfer matrix in Laplace domain is given as $\mathbf{H}(\underline{s}) = \frac{1}{2\underline{s}^4 + 3\underline{s}^3 - \underline{s}^2 + \underline{s} + 1}\begin{pmatrix} 4\underline{s}^2 - 2\underline{s} + 3 \\ \underline{s} - 3 \end{pmatrix}$. Using MATLAB program "**tf2ss**", calculate the matrices **A**, **B**, **C**, and **D** of the system's state-space model. Now, return to equations (1.166) and (1.167) and verify your results.

1.8 Use the Cayley-Hamilton theorem to calculate the state transition matrix $\mathbf{\Phi}(t)$ of matrix $\mathbf{A} = \begin{pmatrix} 1 & 4 \\ 2 & 3 \end{pmatrix}$ in the form of a finite sum.

1.9 Consider the following difference equation of a discrete-time SISO system: $y(k)+2y(k-2) = 3x(k)-2x(k-1)+x(k-2)$.

(a) Find the transfer function $\underline{H}(\underline{z})$ of the system.
(b) Assuming zero initial state conditions, convert the transfer function into state equations and the output equation of an equivalent state-space model.

REFERENCES

[1] T. T. Ha. *Solid-State Microwave Amplifier Design*. John Wiley & Sons, Inc., New York, N.Y., 1981.

[2] S. S. Soliman and M. D. Srinath, *Continuous and Discrete Signals and Systems*. Prentice Hall, Englewood Cliffs, N. J., 1990.

[3] R. D. Strum and D. E. Kirk. *Contemporary Linear Systems - Using MATLAB*. PWS Publishing Company, Boston, MA, 1994.

[4] M. S. Santina, A. R. Stubberud, and G. H. Hostetter. *Digital Control System Design*. Saunders College Publishing, Orlando, FL, 2nd edition, 1994.

[5] The MathWorks, Inc. *Signal Processing Toolbox User's Guide*. Natick, MA, Version 6.0, 2002.

[6] The MathWorks, Inc. *Control System Toolbox User's Guide*. Natick, MA, Version 6.0, 2002.

There are numerous excellent books on time-continuous and time-discrete signals and systems. Below is a concise list of recommended reading material.

[7] R. E. Ziemer, W. H. Tranter, D. R. Fannin. *Signals and Systems: Continuous and Discrete*. Macmillan, New York, N.Y., 2nd edition, 1989.

[8] A. V. Oppenheim and R. W. Schafer. *Discrete-Time Signal Processing*. Prentice Hall, Inc., Englewood Cliffs, N. J., 1989.

[9] J. G. Proakis, D. G. Manolakis. *Introduction to Digital Signal Processing*. Macmillan, New York, N.Y., 1988.

[10] T. Kailath. *Linear Systems*. Prentice Hall, Inc., Englewood Cliffs, N. J., 1980.

[11] D. E Scott. *An Introduction to Circuit Analysis: A Systems Approach*. McGraw-Hill, New York, N.Y., 1987.

[12] R. A. Gabel and R. A. Roberts. *Signals and Linear Systems*. John Wiley & Sons, Inc., New York, N.Y., 3rd edition, 1987.

[13] R. Priemer. *Introductory Signal Processing*. World Scientific Publishing Co., Singapore, 1991.

[14] C. T. Chen. *Linear System Theory and Design*. Holt, Rinehart, and Winston, Inc., New York, N.Y. 1984.

[15] A. V. Oppenheim, A. S. Willsky, and I. T. Young. *Signals and Systems.* Prentice Hall, Inc., Englewood Cliffs, N. J., 1983.

For a nice introduction to MATLAB the reader should consult

[16] L. L. Scharf and R. T. Behrens. *A First Course in Electrical and Computer Engineering, with MATLAB® Programs and Experiments.* Addison-Wesley Publishing Company, Inc., Reading, Mass. 1990.

[17] M. Marcus. *Matrices and MATLAB™ - a Tutorial.* Prentice Hall, Inc., Englewood Cliffs, N. J., 1993.

Chapter 2

ANALYSIS OF SPACE-TIME SIGNALS

2.1 INTRODUCTION TO SPACE-TIME PROCESSES

So far, we have considered a number of MIMO system models with time being the only free parameter. In this chapter, these basic concepts will be extended to signals and systems with more than one independent variable.

First, in Section 2.2, we shall develop the analytical framework for multidimensional signals in time-domain with emphasis on spatial coordinates representing sets of additional free parameters. As we consider information carrying electromagnetic waves, these propagating waves are described as functions of space *and* time. In mechanics, space is the geometric region occupied by bodies [1]. Positions in space are determined relative to some geometric reference system by means of linear and angular measurements. We shall see how these coordinate systems can be appropriately chosen depending on geometrical details of the problem. Vector analysis and differential operators in orthogonal coordinate systems are briefly reviewed and applied to the wave equations.

Then, in Section 2.3, we shall investigate electromagnetic waves and their behavior in linear media. Propagation effects will be considered. In Section 2.4, frequency domain representations of propagating plane waves are introduced. Three additional frequency variables are related to the coordinates in a chosen fixed coordinate system. In addition to the well-known temporal frequency, we shall characterize a propagating wave by its three spatial frequency components, represented in terms of so-called wavenumber vectors.

2.2 CHOICE OF COORDINATE SYSTEM AND VECTOR ANALYSIS

We have already seen that transmission of signals over a general MIMO channel arrangement involves N input ports at the transmit or input side and M output ports or receiver terminals. Now, as we narrow our attention to microwave channels, let us assume that a geometrically fixed array of transmit antennas emit electromagnetic waves propagating through a homogeneous transmission medium. Depending on the individual distances between the transmit and the receive elements, the received signals are filtered versions of the original input signals. Besides these deterministic filtering effects, possible sources of signal degradation include channel noise, multipath fading, reflection, diffraction and scattering of waves, etc. To separate these two categories of influences on received signal shapes, we shall first consider microwave channels with ideal transmission properties and plane waves arriving at the receive antennas. Even under these idealistic assumptions, the received signals must be described as functions of time *and* of their respective distances from the transmit antennas. Thus, signals in time domain are no longer functions of time only. We need a formal description method that includes spatial coordinates as free parameters. Then, in frequency domain, the simple concept of one-dimensional Fourier transform with a single temporal frequency component cannot adequately cover the spatial degrees of freedom of transfer functions.

Location is a three-dimensional quantity. We often need to solve problems where a judicious choice of the origin and the spatial coordinate system will simplify the expressions and the complexity of the problem. The most prominent ones are shown in Figure 2.1. In particular, we can select one of the following standard coordinate systems:

- rectangular or cartesian (x, y, z),
- cylindrical (radius ρ, angle Φ, and length z), or
- spherical or polar (range r, elevation θ, and azimuth Φ).

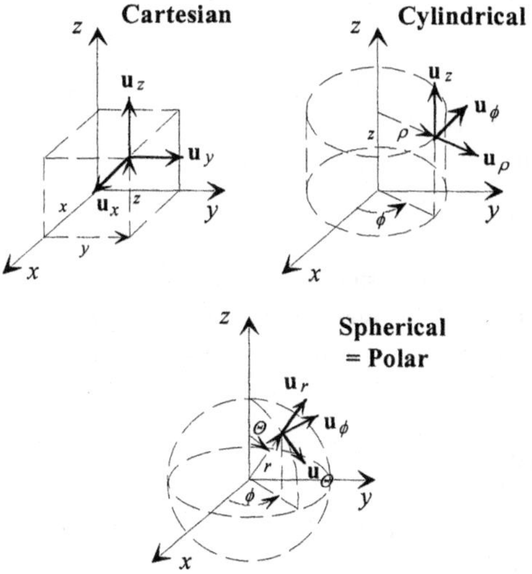

Figure 2-1. Standard coordinate systems and their unit vectors.

Note that each one of them has its particular set of unit vectors, i.e., ($\mathbf{u}_x$, $\mathbf{u}_y$, $\mathbf{u}_z$) for the cartesian system, ($\mathbf{u}_\rho$, $\mathbf{u}_\Phi$, $\mathbf{u}_z$) corresponds to cylindrical coordinates, and ($\mathbf{u}_r$, $\mathbf{u}_\Phi$, $\mathbf{u}_\Theta$) in case of polar coordinates. Spherical (or polar) coordinates are range r, azimuth angle Φ, and elevation angle θ. Azimuth angles Φ are measured relative to the positive x-axis in the rectangular (cartesian) system. Elevation angles θ are measured relative to the positive z-axis. It is sometimes more convenient to use the angle $\bar{\Theta} = \pi/2 - \Theta$ relative to the x,y-plane. Conversion formulas for transformations of vector components are given in Table 2-1.

Table 2-1. Conversion of vector components between standard coordinate systems.

Into cartesian format:

$$\begin{pmatrix} A_x \\ A_y \\ A_z \end{pmatrix} = \begin{pmatrix} \cos(\Phi) & -\sin(\Phi) & 0 \\ \sin(\Phi) & \cos(\Phi) & 0 \\ 0 & 0 & 1 \end{pmatrix} \cdot \begin{pmatrix} A_\rho \\ A_\Phi \\ A_z \end{pmatrix}$$

$$= \begin{pmatrix} \sin(\Theta)\cos(\Phi) & \cos(\Theta)\cos(\Phi) & -\sin(\Phi) \\ \sin(\Theta)\sin(\Phi) & \cos(\Theta)\sin(\Phi) & \cos(\Phi) \\ \cos(\Theta) & -\sin(\Theta) & 0 \end{pmatrix} \cdot \begin{pmatrix} A_r \\ A_\Theta \\ A_\Phi \end{pmatrix}$$

Into cylindrical format:

$$\begin{pmatrix} A_\rho \\ A_\Phi \\ A_z \end{pmatrix} = \begin{pmatrix} \cos(\Phi) & \sin(\Phi) & 0 \\ -\sin(\Phi) & \cos(\Phi) & 0 \\ 0 & 0 & 1 \end{pmatrix} \cdot \begin{pmatrix} A_x \\ A_y \\ A_z \end{pmatrix} = \begin{pmatrix} \sin(\Theta) & \cos(\Theta) & 0 \\ 0 & 0 & 1 \\ \cos(\Theta) & -\sin(\Theta) & 0 \end{pmatrix} \cdot \begin{pmatrix} A_r \\ A_\Theta \\ A_\Phi \end{pmatrix}$$

Into spherical (or polar) format:

$$\begin{pmatrix} A_r \\ A_\Theta \\ A_\Phi \end{pmatrix} = \begin{pmatrix} \sin(\Theta)\cos(\Phi) & \sin(\Theta)\sin(\Phi) & \cos(\Phi) \\ \cos(\Theta)\cos(\Phi) & \cos(\Theta)\sin(\Phi) & -\sin(\Phi) \\ -\sin(\Theta) & \cos(\Theta) & 0 \end{pmatrix} \cdot \begin{pmatrix} A_x \\ A_y \\ A_z \end{pmatrix}$$

$$= \begin{pmatrix} \sin(\Theta) & 0 & \cos(\Theta) \\ \cos(\Theta) & 0 & -\sin(\Theta) \\ 0 & 1 & 0 \end{pmatrix} \cdot \begin{pmatrix} A_\rho \\ A_\Phi \\ A_z \end{pmatrix}$$

Now suppose a measurable scalar quantity f is given as a function of each point (see Figure 2-2) in some region V of space, i.e., $f(\mathbf{p}) = f((x, y, z)^T)$. At one point defined by vector $\mathbf{p}_0 = (x_0, y_0, z_0)^T$, we measure $f(\mathbf{p}_0) = f((x_0, y_0, z_0)^T)$. At another point $\mathbf{p}_0+d\mathbf{p}$, removed by an infinitesimally small distance $d\mathbf{p}$, we obtain $f((x_0+dx, y_0+dy, z_0+dz)^T)$.

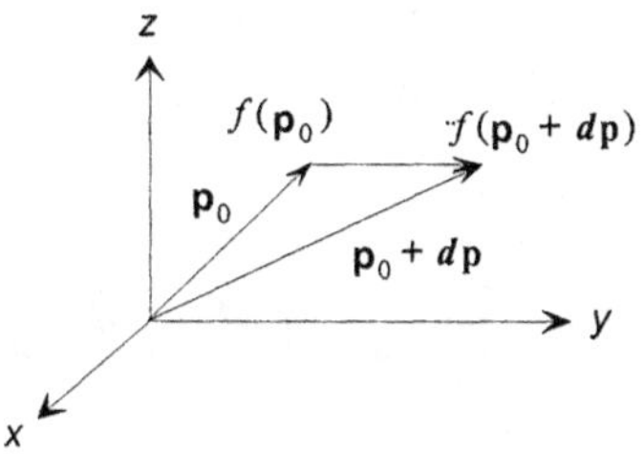

Figure 2-2. Explication of gradient of scalar f in cartesian coordinate system.

Using vector differential calculus, the function value at point $\mathbf{p}_0+d\mathbf{p}$ can be expressed as

$$f((x_0+dx, y_0+dy, z_0+dz)^T) = f((x_0, y_0, z_0)^T) + \left\{ \left(\frac{\partial f}{\partial x}\right)\Big|_{\mathbf{p}_0} dx + \left(\frac{\partial f}{\partial y}\right)\Big|_{\mathbf{p}_0} dy + \left(\frac{\partial f}{\partial z}\right)\Big|_{\mathbf{p}_0} dz \right\} + \cdots. \tag{2.1}$$

Neglecting all higher order differentials, we can write the scalar function increment df between the two points $\mathbf{p}_0$ and $\mathbf{p}_0+d\mathbf{p}$ in the form of

$$df = f(x_0+dx, y_0+dy, z_0+dz) - f(x_0, y_0, z_0) = (\nabla f)\cdot d\mathbf{p} \tag{2.2}$$

where the **gradient** $\nabla f = \text{grad}(f)$ is a vector-valued function defined by

$$\nabla f = \frac{\partial f}{\partial x}\mathbf{u}_x + \frac{\partial f}{\partial y}\mathbf{u}_y + \frac{\partial f}{\partial z}\mathbf{u}_z. \tag{2.3}$$

The directional derivative of a function $f(x,y,z)$ in any direction is the component of ∇f in that direction. Furthermore, $f(x,y,z)$ increases most rapidly in the direction ∇f, and its rate of change in this direction is $|\nabla f|$.

When $\mathbf{F}(x,y,z) = f_x(x,y,z)\mathbf{u}_x+f_y(x,y,z)\mathbf{u}_y+f_z(x,y,z)\mathbf{u}_z$ is a vector-valued function with first partial derivatives, its **divergence** div($\mathbf{F}$) and **curl** (curl($\mathbf{F}$)) are defined as

$$\mathrm{div}(\mathbf{F}) = \nabla \cdot \mathbf{F} = \frac{\partial f_x}{\partial x} + \frac{\partial f_y}{\partial y} + \frac{\partial f_z}{\partial z}, \tag{2.4}$$

$$\mathrm{curl}(\mathbf{F}) = \nabla \times \mathbf{F} = \left(\frac{\partial f_z}{\partial y} - \frac{\partial f_y}{\partial z} \right) \mathbf{u}_x + \left(\frac{\partial f_x}{\partial z} - \frac{\partial f_z}{\partial x} \right) \mathbf{u}_y + \left(\frac{\partial f_y}{\partial x} - \frac{\partial f_x}{\partial y} \right) \mathbf{u}_z$$

$$= \begin{vmatrix} \mathbf{u}_x & \mathbf{u}_y & \mathbf{u}_z \\ \dfrac{\partial}{\partial x} & \dfrac{\partial}{\partial y} & \dfrac{\partial}{\partial z} \\ f_x & f_y & f_z \end{vmatrix} \tag{2.5}$$

Similarly, div(f), div(**F**), and curl(**F**) can be expressed in terms of cylindrical or spherical vector components. An overview of vector differential operators and calculus in orthogonal coordinate systems is given in Appendix A. These operators are essential mathematical tools for the analysis and physical description of electromagnetic phenomena. The choice of coordinates is being made according to symmetry and boundary conditions.

2.3 ELECTROMAGNETIC WAVES PROPAGATING THROUGH MIMO CHANNELS

Let us start our exploration of electromagnetic waves propagating through MIMO channels by writing down a few fundamental equations. These formulas are subsumed in classical physics under the term **Maxwell's equations**. They apply at all times t and for every point $\mathbf{p}$ in space. These coupled partial differential equations are

$$\nabla \times \mathbf{E} = -\frac{\partial \mathbf{B}}{\partial t}, \tag{2.6}$$

$$\nabla \times \mathbf{H} = \frac{\partial \mathbf{D}}{\partial t} + \mathbf{J}, \tag{2.7}$$

$$\nabla \cdot \mathbf{D} = \rho, \tag{2.8}$$

$$\nabla \cdot \mathbf{B} = 0, \tag{2.9}$$

$$\nabla \cdot \mathbf{J} = -\frac{\partial \rho}{\partial t}, \tag{2.10}$$

$$\mathbf{D} = [\varepsilon] \cdot \mathbf{E}, \tag{2.11}$$

$$\mathbf{B} = [\mu] \cdot \mathbf{H} \tag{2.12}$$

where

E = vector of electric field intensity (voltage/distance, V/m),
H = vector of magnetic field intensity (current/distance, A/m),
D = vector of electric flux density (charge/area, As/m^2),
B = vector of magnetic flux density (voltage × time/area, Vs/m^2),
J = vector of electric current density (current/area, A/m^2),
ρ = electric charge density (charge/volume, As/m^3),
$[\varepsilon]$ = permittivity tensor (capacitance/distance, farads/meter = As/(Vm)),
$[\mu]$ = permeability tensor (inductance/distance,
henries/meter = Vs/(Am)).

For a homogeneous isotropic **linear medium** — usually termed a "simple" medium [2] — the constitutive equations (2.11) and (2.12) are

$$\mathbf{D}(\mathbf{p}, t) = \varepsilon \mathbf{E}(\mathbf{p}, t) \tag{2.13}$$

$$\mathbf{B}(\mathbf{p}, t) = \mu \mathbf{H}(\mathbf{p}, t) \tag{2.14}$$

where scalar quantities ε and μ are the permittivity in farads/meter (F/m) and the permeability in henries/meter (H/m), respectively. In free space, the **permittivity** becomes $\varepsilon = \varepsilon_0 = 8.854 \cdot 10^{-12}$ F/m $\approx 1/(36\pi)\ 10^{-9}$ F/m, and the **permeability** is $\mu = \mu_0 = 4\pi \cdot 10^{-7}$ H/m. Lossy linear media are usually characterized by complex-valued scalars

$$\varepsilon = \varepsilon' - j\varepsilon'' = \varepsilon_r \varepsilon_0 (1 - j\,tan(\delta)) \;, \tag{2.15}$$

$$\mu = \mu' - j\mu'' = \mu_0 \frac{1}{(1+\chi_m)} \approx (1+\chi_m)\mu_0 = \mu_r \mu_0 \,. \tag{2.16}$$

In (2.15), the **loss tangent** of a material with conductivity σ is given by

$$tan(\delta) = \frac{\omega\varepsilon'' + \sigma}{\omega\varepsilon'} \,. \tag{2.17}$$

Losses $\omega\varepsilon''$ at a given frequency are due to dielectric damping. ε_r is the dielectric constant.

In (2.16), the complex magnetic susceptibility χ_m of a linear medium accounts for losses due to damping forces produced by **magnetic polarization** $\mathbf{P}_{\mathrm{m}} = \chi_m \mathbf{H}$. We call μ_r the material's relative permeability constant.

In anisotropic dielectrics, such as crystals or ionized gases, the vector of **electric displacement D** is related to the vector of electric field intensity **E** by a tensor permittivity [ε] of rank two (dyad). In cartesian coordinates, we rewrite (2.11) in the form of

$$\begin{pmatrix} D_x \\ D_y \\ D_z \end{pmatrix} = \begin{pmatrix} \varepsilon_{x,x} & \varepsilon_{x,y} & \varepsilon_{x,z} \\ \varepsilon_{y,x} & \varepsilon_{y,y} & \varepsilon_{y,z} \\ \varepsilon_{z,x} & \varepsilon_{z,y} & \varepsilon_{z,z} \end{pmatrix} \cdot \begin{pmatrix} E_x \\ E_y \\ E_z \end{pmatrix} . \tag{2.18}$$

In anisotropic magnetic materials, such as ferrites, the vector of **magnetic induction B** is related to the vector of magnetic field intensity **H** by a tensor permeability [μ] of rank two. Restating (2.12) in rectilinear cartesian coordinates, we may write

$$\begin{pmatrix} B_x \\ B_y \\ B_z \end{pmatrix} = \begin{pmatrix} \mu_{x,x} & \mu_{x,y} & \mu_{x,z} \\ \mu_{y,x} & \mu_{y,y} & \mu_{y,z} \\ \mu_{z,x} & \mu_{z,y} & \mu_{z,z} \end{pmatrix} \cdot \begin{pmatrix} H_x \\ H_y \\ H_z \end{pmatrix} . \tag{2.19}$$

Note that in isotropic linear media the tensors [ε] and [μ] given by (2.18) and (2.19) reduce to diagonal matrices with elements ε and μ, respectively.

Propagation of electromagnetic waves in media — as compared to free space — is governed by interactions of charged particles with electromagnetic fields. At a given point **p** in space and for an operating radian frequency, ω, these effects are best described in the *time* domain by the following three **constitutive vector equations**:

$$\mathbf{D}(\mathbf{p}, t) = \varepsilon(\mathbf{p}, \mathbf{E}, t) * \mathbf{E}(\mathbf{p}, t), \tag{2.20}$$

$$\mathbf{B}(\mathbf{p}, t) = \mu(\mathbf{p}, \mathbf{H}, t) * \mathbf{H}(\mathbf{p}, t), \tag{2.21}$$

$$\mathbf{J}(\mathbf{p}, t) = \sigma(\mathbf{p}, \mathbf{E}, t) * \mathbf{E}(\mathbf{p}, t) \tag{2.22}$$

where * designates the convolution operator and σ stands for the conductivity of the medium in siemens/meter (S/m) or mhos/meter. In free space, we have no losses, hence $\sigma = 0$. All three constitutive parameters ε, μ, and σ may be time-varying quantities. At a distinct position **p** within some media, they may even *nonlinearly* depend on the strength (= magnitude) of the applied vector fields **E** or **H**, respectively. In contrast to these **nonlinear media**, all the others whose constitutive parameters are independent of $|\mathbf{E}|$ and $|\mathbf{H}|$ are termed **linear** media. Yet another important fact to note is that the convolution operators in the constitutive equations reduce to simple products if:

(1) we write down the constitutive equations (2.20 - 2.22) in frequency domain,
(2) if the constitutive parameters don't vary over frequency, or
(3) if the constitutive parameters ε, μ, and σ are constant over time t.

The reader is referred to Chapter 2 of [3] for an excellent overview of electrical properties of matter.

Maxwell's equations constitute a set of *coupled* partial differential equations (PDE's). What we actually need, however, are the *decoupled* field terms $\mathbf{E}(\mathbf{p}, t)$ and $\mathbf{H}(\mathbf{p}, t)$. In a "simple" medium with permittivity ε and permeability μ, we start this decoupling process by taking the curl of (2.6) and (2.7). It follows from (2.8) and (2.11) that

$$\nabla \cdot \mathbf{E} = (1/\varepsilon)\rho. \tag{2.23}$$

Knowing further that the **Laplacian** ∇^2 of a vector function **F** is

$$\nabla^2 \mathbf{F} = \nabla(\nabla \cdot \mathbf{F}) - \nabla \times (\nabla \times \mathbf{F}), \tag{2.24}$$

we obtain

$$\nabla \times (\nabla \times \mathbf{E}) = \nabla (\nabla \cdot \mathbf{E}) - \nabla^2 \mathbf{E} = (1/\varepsilon)\cdot \nabla \rho - \nabla^2 \mathbf{E}. \tag{2.25}$$

Gauss's law for magnetism in a homogeneous medium states that

$$\nabla \cdot \mathbf{B} = \nabla \cdot (\mu \mathbf{H}) = \mu \nabla \cdot \mathbf{H} = 0 \tag{2.26}$$

or

$$\nabla \cdot \mathbf{H} = 0. \tag{2.27}$$

Thus, we have

$$\nabla \times (\nabla \times \mathbf{H}) = \nabla (\nabla \cdot \mathbf{H}) - \nabla^2 \mathbf{H} = - \nabla^2 \mathbf{H}. \tag{2.28}$$

Taking the curl of

$$\nabla \times \mathbf{E} = -\mu \frac{\partial \mathbf{H}}{\partial t} \tag{2.29}$$

and substitution into

$$\nabla \times \mathbf{H} = \mathbf{J} + \varepsilon \frac{\partial \mathbf{E}}{\partial t} \tag{2.30}$$

yields

$$\nabla \times (\nabla \times \mathbf{E}(\mathbf{p}, t)) + \varepsilon\mu \frac{\partial^2 \mathbf{E}(\mathbf{p}, t)}{\partial t^2} = -\mu \frac{\partial \mathbf{J}(\mathbf{p}, t)}{\partial t}. \tag{2.31}$$

It is left as an exercise (see Problem 2.1) to show that

$$\nabla \times (\nabla \times \mathbf{H}(\mathbf{p}, t)) + \varepsilon\mu \frac{\partial^2 \mathbf{H}(\mathbf{p}, t)}{\partial t^2} = \nabla \times \mathbf{J}(\mathbf{p}, t). \tag{2.32}$$

In a **source-free**, **lossy** and **homogeneous** medium with constant conductivity σ, we may replace current density $\mathbf{J}(\mathbf{p}, t)$ by $\sigma\mathbf{E}(\mathbf{p}, t)$. Partial differential equation (2.31) can then be restated in terms of the electric field vectors $\mathbf{E}$ by

$$\nabla^2 \mathbf{E}(\mathbf{p}, t) = \varepsilon\mu \frac{\partial^2 \mathbf{E}(\mathbf{p},t)}{\partial t^2} + \sigma\mu \frac{\partial \mathbf{E}(\mathbf{p},t)}{\partial t}. \tag{2.33}$$

Similarly, since $\nabla \times (\nabla \times \mathbf{H}(\mathbf{p}, t)) = -\nabla^2 \mathbf{H}(\mathbf{p}, \mathrm{t})$ (see (2.28)) and

$$\nabla \times \mathbf{J}(\mathbf{p}, t) = -\sigma\mu \frac{\partial \mathbf{H}(\mathbf{p},t)}{\partial t}, \tag{2.34}$$

equation (2.32) can be written in terms of magnetic field vector **H** and its time derivatives as

$$\nabla^2 \mathbf{H}(\mathbf{p}, t) = \varepsilon\mu \frac{\partial^2 \mathbf{H}(\mathbf{p},t)}{\partial t^2} + \sigma\mu \frac{\partial \mathbf{H}(\mathbf{p},t)}{\partial t}. \tag{2.35}$$

Obviously, in **lossless** homogeneous media, i.e., those with conductivity $\sigma = 0$, the second terms on the right-hand sites of (2.33) and (2.35), respectively, vanish. Equations (2.33) and (2.35) are termed **vector wave equations**. Each of them can be decomposed into three partial differential equations — one per vector component of the electric and the magnetic field.

In **linear** media, the physically and mathematically most relevant solutions to Maxwell's equations (including the vector wave equations!) are steady-state **time-harmonic field** components [4] that satisfy

$$\frac{\partial \mathbf{F}}{\partial t} = j\omega \mathbf{F} \tag{2.36}$$

and

$$\frac{\partial^2 \mathbf{F}}{\partial t^2} = -\omega^2 \mathbf{F}. \tag{2.37}$$

These time translation eigenfunctions or **normal modes** have the form of a real function of space and time. That is, for a given point **p** in an orthogonal coordinate system, we denote a normal mode by

$$f(\mathbf{p}, t) = Re\{f(\mathbf{p}) e^{j\omega t}\}. \tag{2.38}$$

$f(\mathbf{p})$ is the **spatial amplitude profile** that satifies the partial differential equation under consideration. Time-harmonic electromagnetic fields are thus special cases of general time-varying fields. In particular, (2.33) and (2.35) then assume the form of two *inhomogeneous* **Helmholtz equations**

$$\nabla^2 \mathbf{E}(\mathbf{p}, t) = -\omega^2 \varepsilon\mu \mathbf{E}(\mathbf{p}, t) + j\omega\sigma\mu \mathbf{E}(\mathbf{p}, t) = \underline{\gamma}^2 \mathbf{E}(\mathbf{p}, t) \tag{2.39}$$

and

$$\nabla^2 \mathbf{H}(\mathbf{p}, t) = -\omega^2 \varepsilon\mu \mathbf{H}(\mathbf{p}, t) + j\omega\sigma\mu \mathbf{H}(\mathbf{p}, t) = \underline{\gamma}^2 \mathbf{H}(\mathbf{p}, t) \tag{2.40}$$

where $\underline{\gamma} = \alpha + j\beta$ is the **propagation constant.**

Variables α and β are the **attenuation constant** in nepers (Np) per meter and the **phase constant** in radians (rad) per meter, respectively. They are given by

$$\alpha = \omega\sqrt{\varepsilon\mu}\sqrt{\frac{1}{2}\sqrt{1+\left(\frac{\sigma}{\omega\varepsilon}\right)^2} - 1} \qquad \text{Np/m} \tag{2.41}$$

and

$$\beta = \omega\sqrt{\varepsilon\mu}\sqrt{\frac{1}{2}\sqrt{1+\left(\frac{\sigma}{\omega\varepsilon}\right)^2} + 1} \qquad \text{rad/m.} \tag{2.42}$$

It is often required to express the attenuation constant α in decibels (dB) per meter instead of nepers per meter. This is readily done by division by $20\log_{10}(e) \approx 8.686$. In a source-free lossless medium ($\sigma = 0$), there is no attenuation, i.e., $\alpha = 0$. Hence, the time-harmonic electromagnetic field equations (2.39) and (2.40) reduce from inhomogeneous to *homogeneous* Helmholtz equations

$$\nabla^2 \mathbf{E}(\mathbf{p}, t) = -\omega^2 \varepsilon\mu \mathbf{E}(\mathbf{p}, t) = -\beta^2 \mathbf{E}(\mathbf{p}, t) \tag{2.43}$$

and

$$\nabla^2 \mathbf{H}(\mathbf{p}, t) = -\omega^2 \varepsilon\mu \mathbf{H}(\mathbf{p}, t) = -\beta^2 \mathbf{H}(\mathbf{p}, t) \tag{2.44}$$

2.4 WAVENUMBERS AND WAVENUMBER-FREQUENCY SPECTRA

The phase constant β in free space is frequently denoted by the magnitude of the 3D wavenumber vector $\mathbf{k} = (k_{x_1}, k_{x_2}, k_{x_3})^T$ in a generalized orthogonal coordinate system with coordinates x_1, x_2, and x_3. Hence, we can write the square of β as

$$\beta^2 = \omega^2 \varepsilon\mu = |\mathbf{k}|^2 = k_{x_1}^2 + k_{x_2}^2 + k_{x_3}^2 \,. \tag{2.45}$$

The components k_{x_1}, k_{x_2}, and k_{x_3} of **wavenumber vector k** represent the number of waves per unit distance in each of the three orthogonal spatial directions x_1, x_2, and x_3 [5]. For instance, we would express the wavenumber vector as $\mathbf{k} = (k_x, k_y, k_z)^T$ in rectangular (= cartesian) coordinates, $\mathbf{k} = (k_r, k_\Phi, k_\Theta)^T$ in a cylindrical system, or $\mathbf{k} = (k_\rho, k_\Phi, k_z)^T$ in spherical (= polar) coordinates. There are at least 11 of these orthogonal 3D coordinate systems known [4]. In many relevant cases, the vector wave equations can be decoupled into a set of *scalar* Helmholtz equations (e.g., for the electrical field vector **E**) and then solved separately using the most appropriate coordinate system — one that can be most conveniently associated with the given geometry, possible symmetries, and boundary-values of the problem.

Various mathematical methods are known for the analysis of electromagnetic wave problems [5]. Numerical approaches to arbitrary geometries and structures include, among others, the finite difference method (FDM), the finite element method (FEM), the finite integration method (FIM), the boundary element method (BEM), and the Stratton-Chu method [6], [7]. An efficient and accurate software package for modeling of high frequency devices that is widely used in industry is the CST Microwave Studio® [8]. It incorporates the proprietary perfect boundary approximation (PBA) method that allows for accurate modeling of curved surfaces while maintaining the excellent performance of the finite integration method (FIM) in time domain. Since FIM is basically a one-to-one mapping of Maxwell's equations onto a given physical problem in 3D space, it doesn't require any simplifications or approximations. PBA is also a perfect tool to calculate S parameters over large frequency ranges. It works well for all types of antennas, waveguides, microstrip and stripline structures, etc.

To find closed-form solutions to a given electromagnetic wave problem by solving Maxwell's equations directly can become a formidable task. It would by far exceed the scope of this text to deal with the analysis of arbitrary coordinate systems, attenuation due to losses, properties of materials, and layered media — to name just a few of the standard problems encountered. It is, however, instructive to see how the wave equations can be solved in at least three standard orthogonal coordinate systems. We start by taking into account the complex notation of a time-harmonic electric field vector **E** as

$$\begin{aligned}\mathbf{E} &= Re\{|\underline{E}_{x_1}| exp(j\omega t + \varphi_{x_1})\}\mathbf{u}_{x_1} \\ &+ Re\{|\underline{E}_{x_2}| exp(j\omega t + \varphi_{x_2})\}\mathbf{u}_{x_2} \\ &+ Re\{|\underline{E}_{x_3}| exp(j\omega t + \varphi_{x_3})\}\mathbf{u}_{x_3}\end{aligned} \tag{2.46}$$

where E_{xi} (i = 1, 2, 3) is the magnitude, and $\varphi_{xi} = \angle \underline{E}_{xi}$ (i = 1, 2, 3) is the phase of the x_ith vector component. $\mathbf{u}_{xi}$ (i = 1, 2, 3) is the x_ith unit vector in the chosen orthogonal coordinate system, and $Re\{...\}$ stands for the real part of a complex number. Using

$$\mathbf{E} = \underline{E}_{x_1}\mathbf{u}_{x_1} + \underline{E}_{x_2}\mathbf{u}_{x_2} + \underline{E}_{x_3}\mathbf{u}_{x_3}, \tag{2.47}$$

it is apparent that

$$\mathbf{E} = Re\{\mathbf{E}\, exp(j\omega t)\} = \frac{1}{2}(\mathbf{E}\, exp(j\omega t) + \mathbf{E}^*\, exp(-j\omega t)) \tag{2.48}$$

where an asterisk (*) represents the complex conjugate of each of the electrical vector field components. Since we may ignore common phase shifts of all three components, vector **E** is equivalently represented by using its imaginary part $Im\{...\}$ in the form of

$$\mathbf{E} = Im\{\mathbf{E}\, exp(j\omega t)\} = \frac{1}{2j}(\mathbf{E}\, exp(j\omega t) + \mathbf{E}^*\, exp(-j\omega t)). \tag{2.49}$$

Once the electric field vector **E** is found, the components of the magnetic field vector **H** are easily computed in any permissible coordinate system from Faraday's Induction Law (2.29) as

$$\mathbf{H} = \frac{j}{\mu\omega}(\nabla \times \mathbf{E}). \tag{2.50}$$

Coming now to the solutions of the *homogeneous* vector wave equations (source-free, lossless), we take (2.43) and write it in the compact form of

$$\nabla^2(\underline{E}_{x_1}\mathbf{u}_{x_1}+\underline{E}_{x_2}\mathbf{u}_{x_2}+\underline{E}_{x_3}\mathbf{u}_{x_3})-|\mathbf{k}|^2(\underline{E}_{x_1}\mathbf{u}_{x_1}+\underline{E}_{x_2}\mathbf{u}_{x_2}+\underline{E}_{x_3}\mathbf{u}_{x_3})=0. \quad (2.51)$$

The solution method of choice is then to employ the **separation of variables (SoV)** approach which works well if (2.51) can be decoupled into three *independent scalar* partial differential equations. Specifically, if we can express (2.51) as

$$\left.\begin{aligned}\nabla^2\underline{E}_{x_1}(x_1,x_2,x_3)&=-|\mathbf{k}|^2\underline{E}_{x_1}\\ \nabla^2\underline{E}_{x_2}(x_1,x_2,x_3)&=-|\mathbf{k}|^2\underline{E}_{x_2}\\ \nabla^2\underline{E}_{x_3}(x_1,x_2,x_3)&=-|\mathbf{k}|^2\underline{E}_{x_3}\end{aligned}\right\} \quad (2.52)$$

then a solution to any one of these equations, say to the x_ith one (i = 1, 2, 3), is given by the product

$$\underline{E}_{x_i}(x_1,x_2,x_3)=f_1(x_1)f_2(x_2)f_3(x_3) \quad (2.53)$$

where the *i*th scalar function f_i depends on coordinate x_i *only*. Decoupling is possible at least for cartesian and cylindrical coordinates, and closed-form solutions are available for these systems. Unfortunately, in spherical coordinates, we have to assume either **transverse electrical (TE)** or **transverse magnetic (TM) wave** modes [3]. Otherwise, each one of three scalar PDE's includes product terms and/or partial derivatives of range r, azimuth angle Φ, and elevation angle θ. To apply the SoV method, we need an explicit formulation for the Laplacian of a scalar function in the orthogonal coordinate system of interest. For the x_ith electrical field component, this is generally accomplished by calculating

$$\nabla^2\underline{E}_{x_i}=\frac{1}{h_1h_2h_3}\sum_{i=1}^{3}\frac{\partial}{\partial x_i}\left(\frac{h_1h_2h_3}{h_i^2}\frac{\partial\underline{E}_{x_i}}{\partial x_i}\right) \quad (2.54)$$

where

$$h_i = \frac{1}{|\nabla x_i|} = \sqrt{\left(\frac{\partial x}{\partial x_i}\right)^2 + \left(\frac{\partial y}{\partial x_i}\right)^2 + \left(\frac{\partial z}{\partial x_i}\right)^2}, \quad i = 1, 2, 3\,. \tag{2.55}$$

Specifically, it is easy to show that in a **cartesian** system we have

$$h_1 = h_2 = h_3 = 1, \tag{2.56}$$

and thus the Laplacian of the x_ith electrical field component $\underline{E}_{x_i}$ is simply given by

$$\nabla^2 \underline{E}_{x_i} = \frac{\partial^2 \underline{E}_{x_i}}{\partial x^2} + \frac{\partial^2 \underline{E}_{x_i}}{\partial y^2} + \frac{\partial^2 \underline{E}_{x_i}}{\partial z^2}\,. \tag{2.57}$$

In a cartesian system, application of (2.57) to (2.52) yields

$$\left.\begin{aligned} \nabla^2 \underline{E}_x(x,y,z) &= -|\mathbf{k}|^2 \underline{E}_x \\ \nabla^2 \underline{E}_y(x,y,z) &= -|\mathbf{k}|^2 \underline{E}_y \\ \nabla^2 \underline{E}_z(x,y,z) &= -|\mathbf{k}|^2 \underline{E}_z \end{aligned}\right\} \tag{2.58}$$

where each one of these partial differential equations can be solved by an expression of the **product form**

$$\underline{E}_{x_i}(x,y,z) = f_1(x) f_2(y) f_3(z), \qquad Index\; x_i = \{x.\, y, z\}. \tag{2.59}$$

By substituting (2.59) into (2.58), we obtain

$$\begin{aligned} & f_2(y) f_3(z) \frac{\partial^2 f_1(x)}{\partial x^2} + f_1(x) f_3(z) \frac{\partial^2 f_2(y)}{\partial y^2} + f_1(x) f_2(y) \frac{\partial^2 f_3(z)}{\partial z^2} \\ & + |\mathbf{k}|^2 f_1(x) f_2(y) f_3(z) = 0. \end{aligned} \tag{2.60}$$

Replacement of partials by ordinary derivatives and division by $f_1(x)f_2(y)f_3(z)$ yields

$$\frac{1}{f_1(x)}\frac{d^2 f_1(x)}{dx^2}+\frac{1}{f_2(y)}\frac{d^2 f_2(y)}{dy^2}+\frac{1}{f_3(z)}\frac{d^2 f_3(z)}{dz^2}=-|\mathbf{k}|^2 \tag{2.61}$$

which can be written in the form of three separate ordinary differential equations as

$$\left.\begin{aligned}\frac{d^2 f_1(x)}{d^2 x}&=-k_x^2 f_1(x)\\ \frac{d^2 f_2(y)}{d y^2}&=-k_y^2 f_2(y)\\ \frac{d^2 f_3(z)}{d z^2}&=-k_z^2 f_3(z)\end{aligned}\right\} \tag{2.62}$$

under the constraint that the squared magnitude of wavenumber vector **k** must satisfy the **separation equation**

$$|\mathbf{k}|^2=k_x^2+k_y^2+k_z^2 . \tag{2.63}$$

It is easy to verify that valid solutions to (2.62) are, e.g., of the form

$$\left.\begin{aligned}f_1(x)&\propto exp(\pm jk_x x)\\ f_2(y)&\propto exp(\pm jk_y y)\\ f_3(z)&\propto exp(\pm jk_z z)\end{aligned}\right\}. \tag{2.64}$$

Hence, for instance, the instantaneous amplitude of electrical field component E_x can be written as

$$\begin{aligned} E_x(x,y,z,t) &= Re\{\underline{E}_x(x,y,z)exp(j\omega t)\} \\ &= Re\{(a_x^+ exp(-jk_x x) + a_x^- exp(jk_x x) \\ &\quad + a_y^+ exp(-jk_y y) + a_y^- exp(jk_y y) \\ &\quad + a_z^+ exp(-jk_z z) + a_z^- exp(jk_z z))exp(j\omega t)\} \end{aligned} \quad (2.65)$$

with a properly chosen set of coefficients $A = \{a_x^+, a_x^-, a_y^+, a_y^-, a_z^+, a_z^-\}$.

Superscripts + and exponentials with negative signs (e.g., $exp(-jk_x x)$) indicate waves traveling in the positive direction of that axis whereas superscripts - and exponentials with positive signs (e.g., $exp(jk_x x)$) indicate waves traveling in the negative direction. Likewise, we can express instantaneous amplitudes $E_y(x, y, z, t)$ and $E_z(x, y, z, t)$ in forms similar to (2.65) but with different sets of coefficients $B = \{b_x^+, b_x^-, b_y^+, b_y^-, b_z^+, b_z^-\}$ and $C = \{c_x^+, c_x^-, c_y^+, c_y^-, c_z^+, c_z^-\}$.

The constants

$$\left.\begin{aligned} k_x &= |\mathbf{k}|cos(\vartheta_x) \\ k_y &= |\mathbf{k}|cos(\vartheta_y) \\ k_z &= |\mathbf{k}|cos(\vartheta_z) \end{aligned}\right\} \quad (2.66)$$

are termed **wavenumbers** of the **lossless medium** in the x, y, and z direction, respectively. As illustrated in Figure 2-3, direction cosines of a wave are derived from angles ϑ_x, ϑ_y, and ϑ_z between the orientation of wavenumber vector **k** and the positive axes of the rectilinear coordinate system.

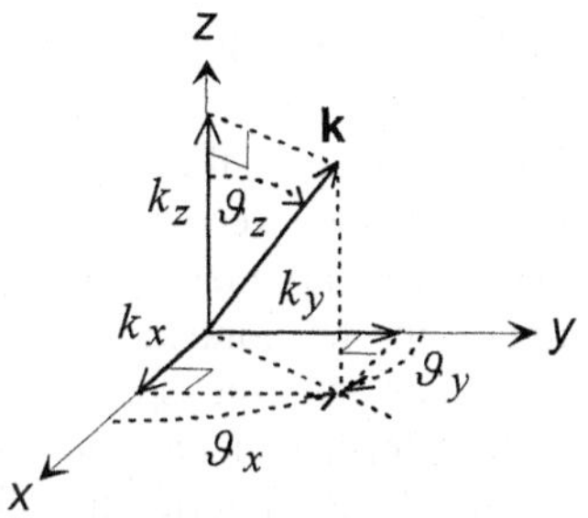

Figure 2-3. Definition of direction cosines and decomposition of wavenumber vector **k** into its cartesian components k_x, k_y, and k_z.

For a wave with wavelength λ and propagation speed c in the direction of unit vector $\mathbf{u}_k$, we note that any given wavenumber vector **k** must satisfy

$$\mathbf{k} = \frac{2\pi}{\lambda}\mathbf{u}_k = \frac{\omega}{c}\mathbf{u}_k \tag{2.67}$$

where $\omega = 2\pi f = 2\pi c/\lambda$ is the radian frequency corresponding to wavelength λ. Obviously, since

$$|\mathbf{k}| = \frac{k_x}{cos(\vartheta_x)} = \frac{k_y}{cos(\vartheta_y)} = \frac{k_z}{cos(\vartheta_z)} = \frac{2\pi}{\lambda} = \frac{\omega}{c}, \tag{2.68}$$

we have

$$\left.\begin{aligned} k_x &= \frac{2\pi}{\lambda}cos(\vartheta_x) = \frac{2\pi}{\lambda_x} \Rightarrow \lambda_x = \frac{\lambda}{cos(\vartheta_x)} \\ k_y &= \frac{2\pi}{\lambda}cos(\vartheta_y) = \frac{2\pi}{\lambda_y} \Rightarrow \lambda_y = \frac{\lambda}{cos(\vartheta_y)} \\ k_z &= \frac{2\pi}{\lambda}cos(\vartheta_z) = \frac{2\pi}{\lambda_z} \Rightarrow \lambda_z = \frac{\lambda}{cos(\vartheta_z)} \end{aligned}\right\} \tag{2.69}$$

For a given radial frequency, say ω_0, or, equivalently, a wave of wavelength λ_0, it is convenient to introduce **elemental instantaneous signals** of the form

$$e(x,y,z,t) = e(\mathbf{p},t) = exp(j(\omega_0 t - \mathbf{k}^T \cdot \mathbf{p})) \qquad (2.70)$$

where the second term in the exponent is the dot product of the transpose of wavenumber vector **k** and the position vector **p** expressed in a cartesian (or any other) coordinate system. There are many good reasons to use the above notation of elemental signals in the context of wave propagation. Most importantly, they open the door into the spectral domain with its four-dimensional (4D) Fourier transform concept and spatio-temporal filtering options. Recalling the 'classical' 1D Fourier transform pair of a real or complex-valued temporal signal $\underline{s}(t)$ and its spectrum $\underline{S}(\omega)$

$$\underline{S}(\omega) = \int_{-\infty}^{\infty} \underline{s}(t) exp(-j\omega t) dt \quad \Leftrightarrow \quad \underline{s}(t) = \frac{1}{2\pi} \int_{-\infty}^{\infty} \underline{S}(\omega) exp(j\omega t) d\omega, \qquad (2.71)$$

we employ elemental instantaneous signals instead of the simple $exp(\pm j\omega t)$ terms in the integrands of (2.71). This leads us to the definition of a 4D **wavenumber-frequency spectrum** of the form

$$\underline{S}(\mathbf{k},\omega) = \int_{-\infty}^{\infty} \int_{-\infty}^{\infty} \underline{s}(\mathbf{p},t) exp(-j(\omega t - \mathbf{k}^T \cdot \mathbf{p})) d\mathbf{p} dt. \qquad (2.72)$$

The inverse 4D Fourier transform operation is then performed by calculation of space-time signal $\underline{s}(\mathbf{p}, t)$ from its spectrum $\underline{S}(\mathbf{k}, \omega)$ using the integral

$$\underline{s}(\mathbf{p},t) = \frac{1}{(2\pi)^4} \int_{-\infty}^{\infty} \int_{-\infty}^{\infty} \underline{S}(\mathbf{k},\omega) exp(j(\omega t - \mathbf{k}^T \cdot \mathbf{p})) d\mathbf{k}\, d\omega. \qquad (2.73)$$

Before we proceed with applications of spatio-temporal signals, their spectra in the $(\mathbf{k}, \omega)$ domain, and the powerful concept of spatial filtering, let us go back to the original wave equation and see if we can find simplifications in the cartesian as well as other orthogonal coordinate systems. The following examples shall explain a few standard approaches. They can also serve as training units. Plots of TEM field components are easily generated by using MATLAB program "**ex2_1.m**".

EXAMPLE 2-1: There is no loss of generality by considering solutions of the wave equation with only one electrical field component. We may arbitrarily choose E_y as a function of cartesian coordinate z only and let $\underline{E}_x = \underline{E}_z = 0$. Then, since $\underline{E}_y$ doesn't depend on x and y, we have

$$\frac{\partial \underline{E}_y}{\partial x} = \frac{\partial \underline{E}_y}{\partial y} = 0 . \tag{2.74}$$

Now, we write down the induction law (2.50) in cartesian format as

$$\nabla \times \mathbf{E} = \begin{pmatrix} \frac{\partial \underline{E}_z}{\partial y} - \frac{\partial \underline{E}_y}{\partial z} \\ \frac{\partial \underline{E}_x}{\partial z} - \frac{\partial \underline{E}_z}{\partial x} \\ \frac{\partial \underline{E}_y}{\partial x} - \frac{\partial \underline{E}_x}{\partial y} \end{pmatrix} = -j\omega\mu \begin{pmatrix} \underline{H}_x \\ \underline{H}_y \\ \underline{H}_z \end{pmatrix} \tag{2.75}$$

to see that the magnetic field vector **H** has only a non-vanishing x component

$$\underline{H}_x = -\frac{j}{\omega\mu}\frac{\partial \underline{E}_y}{\partial z} \tag{2.76}$$

and $\underline{H}_y = \underline{H}_z = 0$.

By Ampère's Law

$$\nabla \times \mathbf{H} = \begin{pmatrix} \frac{\partial \underline{H}_z}{\partial y} - \frac{\partial \underline{H}_y}{\partial z} \\ \frac{\partial \underline{H}_x}{\partial z} - \frac{\partial \underline{H}_z}{\partial x} \\ \frac{\partial \underline{H}_y}{\partial x} - \frac{\partial \underline{H}_x}{\partial y} \end{pmatrix} = j\omega\varepsilon \begin{pmatrix} \underline{E}_x \\ \underline{E}_y \\ \underline{E}_z \end{pmatrix}, \tag{2.77}$$

we may infer that $\partial \underline{H}_x / \partial y = 0$. Knowing further that

$$\mathrm{div}(\mathbf{H}) = \nabla \cdot \mathbf{H} = \frac{\partial \underline{H}_x}{\partial x} + \frac{\partial \underline{H}_y}{\partial y} + \frac{\partial \underline{H}_z}{\partial z} = 0, \tag{2.78}$$

we also have $\partial \underline{H}_x / \partial x = 0$. Thus, since the two remaining differential equations are

$$\frac{\partial \underline{E}_y}{\partial z} = j\omega\,\mu \underline{H}_x \tag{2.79}$$

and

$$\frac{\partial \underline{H}_x}{\partial z} = j\omega\,\varepsilon\,\underline{E}_y\,, \tag{2.80}$$

we may substitute (2.80) into (2.79) and find the wave equation

$$\frac{\partial^2 \underline{E}_y(z)}{\partial z^2} = -\,\omega^2 \varepsilon\,\mu\,\underline{E}_y(z) = -\,k_z^2\,\underline{E}_y(z) = -\,\beta^2 \underline{E}_y(z) \tag{2.81}$$

which is merely a special form of the homogeneous Helmholtz equations (2.43) in a source-free and lossless medium. Note the special form of wavenumber vector **k** in (2.81) where $k_x = k_y = 0$ and $|\mathbf{k}| = k_z = \beta,\ (k_z, \beta \text{ positive, real})$. Solutions to (2.81) can be considered as superpositions of two waves traveling in opposite directions of the z axis. Using electrical field amplitudes E^+ (+z traveling wave) and E^- (-z traveling wave), we may write

$$\underline{E}_y(z) = E_y^+\,exp(-jk_z z) + E_y^-\,exp(jk_z z)\,. \tag{2.82}$$

The magnetic field is then calculated, using (2.76) and $k_z = \beta = \omega\sqrt{\varepsilon\,\mu}$, as

$$\begin{aligned}
\underline{H}_x &= -\frac{j}{\omega\,\mu}\frac{\partial \underline{E}_y}{\partial z} = -\frac{j}{\omega\,\mu}(-jk_z E_y^+\,exp(-jk_z z) + jk_z E_y^-\,exp(jk_z z)) \\
&= \sqrt{\frac{\varepsilon}{\mu}}(-E_y^+\,exp(-jk_z z) + E_y^-\,exp(jk_z z)) \\
&= -\frac{1}{\eta}E_y^+\,exp(-jk_z z) + \frac{1}{\eta}E_y^-\,exp(jk_z z)
\end{aligned} \tag{2.83}$$

where the ratio

$$\eta = \sqrt{\frac{\mu}{\varepsilon}} \tag{2.84}$$

is usually termed the **intrinsic impedance** of the medium. It has units of ohms (V/A). By taking the ratio of the electric to magnetic field, we may define the **wave impedance**

$$Z_w = -\frac{E_y^+}{H_x^+} = \frac{E_y^-}{H_x^-} = \frac{\omega\,\mu}{\beta} = \eta = \sqrt{\frac{\mu}{\varepsilon}}\,. \tag{2.85}$$

Note that in free space we have

$$\eta = \eta_0 = \sqrt{\frac{\mu_0}{\varepsilon_0}} \approx 120\pi \approx 377 \text{ ohms}. \tag{2.86}$$

In our example, the **E** and **H** vectors are orthogonal to each other and also to the direction of propagation $\pm z$. We call this particular electromagnetic wave a **transverse electromagnetic** or **TEM wave**. It is also a **plane wave** because, over a set of planes defined by $z = const.$, the phase $k_z z$ of $\underline{E}_y$ and $\underline{H}_x$ is constant. The instantaneous field components of a time-harmonic wave traveling in the $+z$- direction are illustrated in Figure 2-4. MATLAB program "**ex2_1.m**" may be run to produce the function plots.

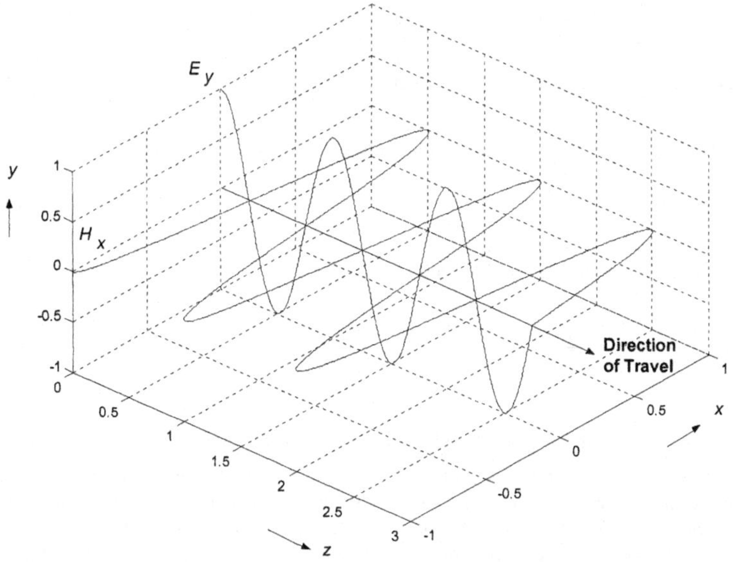

Figure 2-4. Electric and magnetic field components of TEM wave described in Example 2.1.

The real parts of the electric and magnetic field may be written as

$$\begin{aligned} E_y^+(z,t) &= E_0 \cos(\omega t - k_z z) \\ H_x^+(z,t) &= -\frac{E_0}{\eta}\cos(\omega t - k_z z) \end{aligned} \tag{2.87}$$

Phasors of wave components propagating in the $+z$ direction with initial phase angle φ_z^+ are described by their magnitude and phase in the form of

$$\underline{E}^+ = \left|\underline{E}^+\right| exp(j\varphi_z^+) . \tag{2.88}$$

The instantaneous value of a time-harmonic E_y^+ field is thus given as a function of both spatial coordinate z and time t by

$$E_y^+(z,t) = Re\{\left|\underline{E}_y^+\right| exp(-j\beta z) exp(j\omega t + \varphi_z^+)\} = \left|\underline{E}_y^+\right| cos(\omega t - \beta z + \varphi_z^+) . \tag{2.89}$$

If an arbitrarily chosen point on that wave (e.g., the maximum) is observed at slightly different times t and $(t + \Delta t)$ it moves by distance Δz in the positive z direction. To calculate the velocity of that particular marker point, let us assume that there is no dispersion in the medium (as is always the case in free space). Then, by equating instantaneous amplitudes

$$cos(\omega t - \beta z + \varphi_z^+) = cos(\omega(t + \Delta t) - \beta(z + \Delta z) + \varphi_z^+) \tag{2.90}$$

and comparing arguments of the cosine functions on both sides, we have

$$\omega \Delta t - \beta \Delta z = 0 . \tag{2.91}$$

The quotient of Δz and Δt is called **phase velocity**

$$v_p = \frac{\Delta z}{\Delta t} = \frac{\omega}{\beta} = \frac{1}{\sqrt{\varepsilon\mu}} = \frac{c_0}{\sqrt{\varepsilon_r \mu_r}} \tag{2.92}$$

where $c_0 = 1/\sqrt{\varepsilon_0 \mu_0}$ is the speed of light in vacuum. It is interesting to note that, for any non-dispersive medium, v_p is independent of the wave's frequency. It is just a material-dependent constant characterized by the electric permittivity and magnetic permeability of the medium. Phase velocity should, however, not be confused with **group velocity** (the velocity with which energy propagates). Group velocity v_g is related to v_p and the **group delay** τ_g by

$$v_g = \frac{d\omega}{d\beta} = \frac{1}{\tau_g} . \tag{2.93}$$

v_g is readily calculated using the quotient rule for derivatives and taking into account that its reciprocal is

$$\tau_g = \frac{1}{v_g} = \frac{d\beta}{d\omega} = \frac{d(\omega / v_p)}{d\omega} = \frac{v_p - \omega \dfrac{dv_p}{d\omega}}{v_p^2} . \tag{2.94}$$

Hence, we obtain

$$v_g = \frac{v_p}{1 - \dfrac{\omega}{v_p}\dfrac{dv_p}{d\omega}} = \frac{v_p}{1 + \dfrac{\omega}{c_0} v_p \dfrac{dn_r}{d\omega}} \tag{2.95}$$

where

$$n_r = \frac{c_0}{v_p} = c_0 \frac{\beta}{\omega} = \sqrt{\varepsilon_r \mu_r} \tag{2.96}$$

is the medium's **index of refraction** — a physical quantity well-known in optics.

Since, in a **dispersionless medium,** the group velocity is independent of radial frequency ω (i.e., $\mathrm{d}v_p/\mathrm{d}\,\omega = \mathrm{d}n/\mathrm{d}\,\omega = 0$), we may infer that in these media

$$v_p = v_g = \frac{c_0}{\sqrt{\varepsilon_r \mu_r}} \quad \Leftrightarrow \quad v_p v_g = \frac{c_0^2}{\varepsilon_r \mu_r} = const. \tag{2.97}$$

Note that other media, such as all dielectric materials, may of course show dispersive behavior. In particular, we distinguish between media with "normal" dispersion where

$$\frac{dv_p}{d\omega} < 0 \quad \Leftrightarrow \quad \frac{dn_r}{d\omega} = -\frac{c_0}{v_p^2}\frac{dv_p}{d\omega} > 0 \quad \Leftrightarrow \quad v_g > 0 \tag{2.98}$$

and others which, mostly within small characteristic frequency bands, exhibit "anomalous" or "abnormal" dispersion where

$$\frac{dv_p}{d\omega} > 0 \quad \Leftrightarrow \quad \frac{dn_r}{d\omega} = -\frac{c_0}{v_p^2}\frac{dv_p}{d\omega} < 0 \quad \Leftrightarrow \quad v_g < 0\,. \tag{2.99}$$

▶◀

EXAMPLE 2-2: The separation of variables (SoV) technique works well for problems with **cylindrical** geometry. In this example, we shall find solutions to the wave equation specialized to cylindrical coordinates. We start by recalling that, according to (2.54), the Laplacian of the x_i-th electrical field component is to be calculated for cylindrical coordinates $x_i = \{\rho, \Phi, z\}$. From Table 2-1 and equation (2.55) we may infer that $h_1 = h_3 = 1$ and $h_2 = \rho$. Application of

$$\nabla^2 \underline{E}_{x_i} = \frac{1}{h_1 h_2 h_3} \sum_{i=1}^{3} \frac{\partial}{\partial x_i}\left(\frac{h_1 h_2 h_3}{h_i^2} \frac{\partial \underline{E}_{x_i}}{\partial x_i} \right) \quad \text{Index } x_i = \{\rho, \Phi, z\} \tag{2.100}$$

then yields

$$\nabla^2 \underline{E}_{x_i} = \frac{1}{\rho}\frac{\partial}{\partial \rho}(\rho \frac{\partial \underline{E}_{x_i}}{\partial \rho}) + \frac{1}{\rho^2}\frac{\partial^2 \underline{E}_{x_i}}{\partial \Phi^2} + \frac{\partial^2 \underline{E}_{x_i}}{\partial z^2}$$
$$= \frac{\partial^2 \underline{E}_{x_i}}{\partial \rho^2} + \frac{1}{\rho}\frac{\partial \underline{E}_{x_i}}{\partial \rho} + \frac{1}{\rho^2}\frac{\partial^2 \underline{E}_{x_i}}{\partial \Phi^2} + \frac{\partial^2 \underline{E}_{x_i}}{\partial z^2} \tag{2.101}$$

Writing the electrical field vector **E** in the form of

$$\mathbf{E} = \underline{E}_\rho \mathbf{u}_\rho + \underline{E}_\Phi \mathbf{u}_\Phi + \underline{E}_z \mathbf{u}_z \tag{2.102}$$

and substituting it into the wave equation

$$\nabla^2 (\underline{E}_\rho \mathbf{u}_\rho + \underline{E}_\Phi \mathbf{u}_\Phi + \underline{E}_z \mathbf{u}_z) - |\mathbf{k}|^2 (\underline{E}_\rho \mathbf{u}_\rho + \underline{E}_\Phi \mathbf{u}_\Phi + \underline{E}_z \mathbf{u}_z) = 0 \tag{2.103}$$

yields the following system of three scalar partial differential equations

$$\left.\begin{aligned} &\nabla^2 \underline{E}_\rho + (-\frac{1}{\rho^2}\underline{E}_\rho - \frac{2}{\rho^2}\frac{\partial \underline{E}_\Phi}{\partial \Phi}) = -|\mathbf{k}|^2 \underline{E}_\rho \\ &\nabla^2 \underline{E}_\Phi + (-\frac{1}{\rho^2}\underline{E}_\Phi + \frac{2}{\rho^2}\frac{\partial \underline{E}_\rho}{\partial \Phi}) = -|\mathbf{k}|^2 \underline{E}_\Phi \\ &\nabla^2 \underline{E}_z = -|\mathbf{k}|^2 \underline{E}_z \end{aligned}\right\} \tag{2.104}$$

Unlike the cartesian system of (2.58), note that here the first two second-order differential equations are coupled, because $\underline{E}_\rho$ and $\underline{E}_\Phi$ appears in both of them.

Obviously, each one of the equations in (2.104) contains a Laplacian of the form (2.101), and we may separate each electrical field component into a product of three scalar functions, i.e.,

$$\underline{E}_{x_i}(\rho, \Phi, z) = f_1(\rho) f_2(\Phi) f_3(z) \qquad \textit{Index } x_i = \{\rho. \Phi, z\}. \tag{2.105}$$

With a few differential calculus steps (see, e.g., [3] or [4]), it is straightforward to show that (2.104) can be separated as

$$\left.\begin{array}{l} \rho^2 \dfrac{d^2 f_1}{d\rho^2} + \rho \dfrac{df_2}{d\rho} = (\upsilon^2 - (k_\rho \rho)^2) f_1 \\[2ex] \dfrac{d^2 f_2}{d\Phi^2} = -\upsilon^2 f_2 \\[2ex] \dfrac{d^2 f_3}{dz^2} = -k_z f_3 \end{array}\right\} \tag{2.106}$$

where, by definition, the squares of the propagation constants are constrained by

$$k_\rho^2 + k_z^2 = |\mathbf{k}|^2 = \beta^2 \tag{2.107}$$

The first equation in (2.106) is well-known in mathematics as the **Bessel equation.** It is a second-order differential equation with two independent solutions, one that is analytic at f_1 = 0, and one that is singular at f_1 = 0. Solutions of the Bessel equation are called cylinder functions, commonly defined as linear combinations

$$\underline{C}_\upsilon(k_\rho \rho) = a_1 J_\upsilon(k_\rho \rho) + a_2 N_\upsilon(k_\rho \rho), \quad a_1, a_2 \text{ constants} \tag{2.108}$$

or

$$\underline{C}_\upsilon(k_\rho \rho) = b_1 H_\upsilon^{(1)}(k_\rho \rho) + b_2 H_\upsilon^{(2)}(k_\rho \rho), \quad b_1, b_2 \text{ constants}. \tag{2.109}$$

In (2.108) and (2.109), four types of Bessel functions appear. They are characterized by their kind, order υ, and argument ρ.

Special cases of cylinder functions include the **Bessel function of the first kind** and of order υ

$$J_\upsilon(k_\rho \rho) = \left(\frac{k_\rho \rho}{2}\right)^\upsilon \sum_{m=0}^{\infty} \frac{(-1)^m}{m!(\upsilon + m)!} \left(\frac{k_\rho \rho}{2}\right)^{2m}, \quad |\arg k_\rho \rho| < \pi, \tag{2.110}$$

the **Bessel function of the second kind** (= **Neumann function**)

$$N_\upsilon(k_\rho \rho) = \frac{1}{\sin(\upsilon \pi)} \left(J_\upsilon(k_\rho \rho) \cos(\upsilon \pi) - J_{-\upsilon}(k_\rho \rho) \right), \tag{2.111}$$

and the two **Bessel functions of the third kind** (= **Hankel functions**) with superscripts $^{(1)}$ and $^{(2)}$, respectively, defined as

$$\underline{H}_\upsilon^{(1)}(k_\rho \rho) = J_\upsilon(k_\rho \rho) + jN_\upsilon(k_\rho \rho) \tag{2.112}$$

and

$$\underline{H}_{\upsilon}^{(2)}(k_{\rho}\rho) = J_{\upsilon}(k_{\rho}\rho) - jN_{\upsilon}(k_{\rho}\rho)\,. \tag{2.113}$$

Knowing **cylinder functions** $\underline{C}_{\upsilon}(k_{\rho}\rho)$ and $\underline{C}_{\upsilon-1}(k_{\rho}\rho)$ of any kind, it is convenient to calculate the cylinder function of order $\upsilon+1$ by means of recurrence formulas such as

$$\underline{C}_{\upsilon+1}(k_{\rho}\rho) = \frac{2\upsilon}{k_{\rho}\rho}\underline{C}_{\upsilon}(k_{\rho}\rho) - \underline{C}_{\upsilon-1}(k_{\rho}\rho) \tag{2.114}$$

and

$$\underline{C}_{\upsilon+1}(k_{\rho}\rho) = \underline{C}_{\upsilon-1}(k_{\rho}\rho) - 2\frac{d\underline{C}_{\upsilon}(k_{\rho}\rho)}{d(k_{\rho}\rho)}\,. \tag{2.115}$$

These functional relations, series and integral representations, asymptotic expansions, "addition theorems", etc. are well documented, e.g., in [9, 10]. Moreover, *m* files "**besselj.m**" and "**bessely.m**" are included in MATLAB's package of specialized mathematical functions. Bessel functions of the first kind and of the second kind, respectively, are readily computed using these *m* files.

With the exception of the first one of the three differential equation in (2.106), the two others are of the type displayed in equation (2.62). Their solutions $f_2(\Phi)$ and $f_3(z)$ are hence known to be linear combinations of harmonic functions

$$\left.\begin{aligned} f_2(\Phi) &\propto exp(\pm jk_{\Phi}\Phi) \\ f_3(z) &\propto exp(\pm jk_z z) \end{aligned}\right\} \tag{2.116}$$

Therefore, similar to (2.65), we may express the axial *z* component of the electrical field as

$$\begin{aligned} E_z(\rho,\Phi,z,t) &= Re\{\underline{E}_z(\rho,\Phi,z)exp(j\omega t)\} \\ &= Re\{(a_{\rho}^{+}J_{\upsilon}(-jk_{\rho}\rho) + a_{\rho}^{-}N_{\upsilon}(jk_{\rho}\rho) \\ &\quad + a_{\Phi}^{+}exp(-j\upsilon\Phi) + a_{\Phi}^{-}exp(j\upsilon\Phi) \\ &\quad + a_z^{+}exp(-jk_z z) + a_z^{-}exp(jk_z z))exp(j\omega t)\} \end{aligned} \tag{2.117}$$

where $a = \{\, a_{\rho}^{+}, a_{\rho}^{-}, a_{\Phi}^{+}, a_{\Phi}^{-}, a_z^{+}, a_z^{-} \,\}$ is a set of properly chosen coefficients depending on physical constraints of the given boundary-value problem. To find the full set of electromagnetic field components, it is more common to employ **wave potentials** Ψ instead of electrical field components. Then, in a source-free homogeneous region satisfying

$$\begin{aligned} \nabla\times\mathbf{E} &= -j\omega\mu\mathbf{H}, \quad \nabla\cdot\mathbf{H}=0 \\ \nabla\times\mathbf{H} &= \ \ j\omega\varepsilon\mathbf{E}, \quad \nabla\cdot\mathbf{E}=0 \end{aligned} \tag{2.118}$$

the time-harmonic fields are expressed in terms of a **magnetic vector potential A** or an **electric vector potential F**. Dealing with magnetic and electrical vector potentials separately, we may write

$$\begin{aligned} \nabla^2\mathbf{A} + |\mathbf{k}|^2\mathbf{A} &= 0 \\ \nabla^2\mathbf{F} + |\mathbf{k}|^2\mathbf{F} &= 0 \end{aligned} \tag{2.119}$$

It is important to note that the vector potential approach starts by letting the wave potential Ψ be a *rectangular* component (say, the z component) of vector potential **A** or **F**, respectively. Either of the two vector potentials is intentionally set to be zero, and a single *scalar* partial differential equation (Helmholtz Equation) is solved in terms of wave potential Ψ. So, if we choose a **transverse electric** (or **TE**) **field**, i.e., a field with no electrical z component ($\underline{E}_z = 0$), we have

$$\mathbf{A}=\mathbf{0}, \quad \mathbf{F}=\Psi\,\mathbf{u_z}=(0,0,\Psi)^{\mathrm{T}}. \tag{2.120}$$

Once we know appropriate solutions Ψ to the scalar Helmholtz Equation (in the selected coordinate system!), we also know the z component of **F**. The remaining five components of the TE-type electromagnetic field may then be calculated by expanding

$$\mathbf{E}=-\nabla\times\mathbf{F}, \qquad \mathbf{H}=-j\left(\omega\varepsilon\mathbf{F}+\frac{1}{\omega\mu}\nabla(\nabla\cdot\mathbf{F})\right) \tag{2.121}$$

with respect to chosen coordinate system.

Similarly, if we assume a vanishing z component of the magnetic field ($\underline{H}_z = 0$), the resulting electromagnetic field is termed **transverse magnetic** to the z direction or a **TM field**.

Then, by letting the vector potentials be

$$\mathbf{F}=\mathbf{0}, \quad \mathbf{A}=\Psi\,\mathbf{u_z}=(0,0,\Psi)^T, \tag{2.122}$$

the first step must be to find solutions to the Helmholtz Equation. Also, knowing Ψ and **A**, the remaining five electromagnetic field components are obtained by expanding

$$\mathbf{H}=\nabla\times\mathbf{A}, \qquad \mathbf{E}=-j\left(\omega\mu\mathbf{A}+\frac{1}{\omega\varepsilon}\nabla(\nabla\cdot\mathbf{A})\right) \tag{2.123}$$

Writing the Helmholtz Equation in cylindrical coordinates, we have

$$\nabla^2\Psi + |\mathbf{k}|^2\Psi = \frac{1}{\rho}\frac{\partial}{\partial\rho}(\rho\frac{\partial\Psi}{\partial\rho}) + \frac{1}{\rho^2}\frac{\partial^2\Psi}{\partial\Phi^2} + \frac{\partial^2\Psi}{\partial z^2} + |\mathbf{k}|^2\Psi$$
$$= \frac{\partial^2\Psi}{\partial\rho^2} + \frac{1}{\rho}\frac{\partial\Psi}{\partial\rho} + \frac{1}{\rho^2}\frac{\partial^2\Psi}{\partial\Phi^2} + \frac{\partial^2\Psi}{\partial z^2} + |\mathbf{k}|^2\Psi = 0. \quad (2.124)$$

As explained above, solutions to this equation can be found using the SoV method. In particular, elementary wave functions of the type

$$\underline{\Psi} = \underline{C}_{\upsilon}(k_{\rho}\rho)exp(j\upsilon\Phi)exp(jk_z z) \quad (2.125)$$

and linear combinations thereof are possible solutions. Note, however, that these linear combinations must be double sums over k_{ρ} and order υ or, alternatively, k_z and υ. Triple sums or double sums over k_{ρ} and k_z are not allowed because of interrelationship (2.114) that imposes the constraint $k_{\rho}^2 + k_z^2 = |\mathbf{k}|^2$.

Application of (2.121) then yields all vector components of the **TE** electromagnetic field in the cylindrical coordinate system as follows:

$$\begin{aligned} \underline{E}_{\rho} &= -\frac{1}{\rho}\frac{\partial\underline{\Psi}}{\partial\Phi} & \underline{H}_{\rho} &= -j\frac{1}{\omega\mu}\frac{\partial^2\underline{\Psi}}{\partial\rho\,\partial z} \\ \underline{E}_{\Phi} &= \frac{\partial\underline{\Psi}}{\partial\rho} & \underline{H}_{\Phi} &= -j\frac{1}{\omega\mu\,\rho}\frac{\partial^2\underline{\Psi}}{\partial\Phi\,\partial z} \\ \underline{E}_z &= 0 & \underline{H}_z &= -j\frac{1}{\omega\mu}\left(\frac{\partial^2\underline{\Psi}}{\partial z^2} + |\mathbf{k}|^2\underline{\Psi}\right) \end{aligned} \quad (2.126)$$

Finally, if the components of a **TM** electromagnetic field are needed in cylindrical coordinates, we may infer from (2.123) that

$$\begin{aligned} \underline{H}_{\rho} &= \frac{1}{\rho}\frac{\partial\underline{\Psi}}{\partial\Phi} & \underline{E}_{\rho} &= -j\frac{1}{\omega\varepsilon}\frac{\partial^2\underline{\Psi}}{\partial\rho\,\partial z} \\ \underline{H}_{\Phi} &= -\frac{\partial\underline{\Psi}}{\partial\rho} & \underline{E}_{\Phi} &= -j\frac{1}{\omega\varepsilon\,\rho}\frac{\partial^2\underline{\Psi}}{\partial\Phi\,\partial z} \\ \underline{H}_z &= 0 & \underline{E}_z &= -j\frac{1}{\omega\varepsilon}\left(\frac{\partial^2\underline{\Psi}}{\partial z^2} + |\mathbf{k}|^2\underline{\Psi}\right) \end{aligned} \quad (2.127)$$

►◄

EXAMPLE 2-3: In this example, we examine electromagnetic fields in the **spherical (= polar) coordinate system** where any vector is defined by its radius r, elevation angle Θ, and azimuth angle Φ. The electrical field vector may be written in the form of

$$\mathbf{E} = \underline{E}_r \mathbf{u}_r + \underline{E}_\Theta \mathbf{u}_\Theta + \underline{E}_\Phi \mathbf{u}_\Phi , \tag{2.128}$$

and the wave equation in a source-free and lossless medium is given by

$$\nabla^2 (\underline{E}_r \mathbf{u}_r + \underline{E}_\Theta \mathbf{u}_\Theta + \underline{E}_\Phi \mathbf{u}_\Phi) - |\mathbf{k}|^2 (\underline{E}_r \mathbf{u}_r + \underline{E}_\Theta \mathbf{u}_\Theta + \underline{E}_\Phi \mathbf{u}_\Phi) = 0 . \tag{2.129}$$

From Table 2-1 and equation (2.55), we may infer that $h_1 = 1$, $h_2 = r$, and $h_3 = r \sin\Theta$. Application of

$$\nabla^2 \underline{E}_{x_i} = \frac{1}{h_1 h_2 h_3} \sum_{i=1}^{3} \frac{\partial}{\partial x_i} \left(\frac{h_1 h_2 h_3}{h_i^2} \frac{\partial \underline{E}_{x_i}}{\partial x_i} \right) \qquad \mathit{Index}\; x_i = \{r, \Theta, \Phi\} \tag{2.130}$$

yields

$$\begin{aligned} \nabla^2 \underline{E}_{x_i} &= \frac{1}{r^2} \frac{\partial}{\partial r} \left(r^2 \frac{\partial \underline{E}_{x_i}}{\partial r} \right) + \frac{1}{r^2 \sin\Theta} \frac{\partial}{\partial \Theta} \left(\sin\Theta \frac{\partial \underline{E}_{x_i}}{\partial \Theta} \right) + \frac{1}{r^2 \sin^2 \Theta} \frac{\partial^2 \underline{E}_{x_i}}{\partial \Phi^2} = \\ &= \frac{\partial^2 \underline{E}_{x_i}}{\partial r^2} + \frac{2}{r} \frac{\partial \underline{E}_{x_i}}{\partial r} + \frac{1}{r^2} \left(\frac{\partial}{\partial \xi} \left(\left(1 - \xi^2\right) \frac{\partial \underline{E}_{x_i}}{\partial \xi} \right) + \frac{1}{1 - \xi^2} \frac{\partial^2 \underline{E}_{x_i}}{\partial \Phi^2} \right) \end{aligned} \tag{2.131}$$

where $\xi = \cos\Theta$.

Although the wave equation can be written as a set of three partial differential equations in the form of (see reference [3])

$$\begin{aligned} &\nabla^2 \underline{E}_r - \frac{2}{r^2} \left(\underline{E}_r + \underline{E}_\Theta \cot\Theta + \csc\Theta \frac{\partial \underline{E}_\Phi}{\partial \Phi} + \frac{\partial \underline{E}_\Theta}{\partial \Theta} \right) + |\mathbf{k}|^2 \underline{E}_r = 0 \\ &\nabla^2 \underline{E}_\Theta - \frac{1}{r^2} \left(\underline{E}_\Theta \csc^2 \Theta - 2 \frac{\partial \underline{E}_r}{\partial \Theta} + 2 \cot\Theta \csc\Theta \frac{\partial \underline{E}_\Phi}{\partial \Phi} \right) + |\mathbf{k}|^2 \underline{E}_\Theta = 0 \\ &\nabla^2 \underline{E}_\Phi - \frac{1}{r^2} \left(\underline{E}_\Phi \csc^2 \Theta - 2 \csc\Theta \frac{\partial \underline{E}_r}{\partial \Phi} - 2 \cot\Theta \csc\Theta \frac{\partial \underline{E}_\Theta}{\partial \Phi} \right) + |\mathbf{k}|^2 \underline{E}_\Phi = 0 \end{aligned} \tag{2.132}$$

each one of these equations contains expressions of r, Θ, and Φ. Therefore, (2.132) constitutes a set of *coupled* partial differential equations that cannot be solved by means of the SoV method directly. We may, for instance, apply a similar approach as in the cylindrical coordinate system. Since the spherical coordinate system doesn't have a rectangular z component, one option would be to generate the missing z components of **A** or **F** vector potentials by decomposing them into their r and Θ parts. Hence, our calculation of **TM** fields in spherical coordinate systems might start by letting the vector potentials be

$$\mathbf{A} = \Psi \mathbf{u_z} = \Psi(cos(\Theta)\mathbf{u_r} - sin(\Theta)\mathbf{u_\Theta}) \qquad \mathbf{F} = \mathbf{0}, \tag{2.133}$$

then, by means of the SoV method, finding complex-valued solutions

$$\Psi = f_1(r) f_2(\Theta) f_3(\Phi) \tag{2.134}$$

to the scalar Helmholtz Equation

$$\begin{aligned} &\nabla^2\Psi + |\mathbf{k}|^2\Psi = \\ &= \frac{1}{r^2}\frac{\partial}{\partial r}\left(r^2\frac{\partial\Psi}{\partial r}\right) + \frac{1}{r^2 sin\Theta}\frac{\partial}{\partial\Theta}\left(sin\Theta\frac{\partial\Psi}{\partial\Theta}\right) + \frac{1}{r^2 sin^2\Theta}\frac{\partial^2\Psi}{\partial\Phi^2} + |\mathbf{k}|^2\Psi = \\ &= \frac{\partial^2\Psi}{\partial r^2} + \frac{2}{r}\frac{\partial\Psi}{\partial r} + \frac{1}{r^2}\left(\frac{\partial}{\partial\xi}\left(\left(1-\xi^2\right)\frac{\partial\Psi}{\partial\xi}\right) + \frac{1}{1-\xi^2}\frac{\partial^2\Psi}{\partial\Phi^2}\right) + |\mathbf{k}|^2\Psi = 0 \end{aligned} \tag{2.135}$$

where $\xi = cos\,\Theta$. Solutions are of the form

$$\begin{aligned} \underline{\Psi} = \underline{\Psi}_{m,n} &= \left(a_r^+ j_n(kr) + a_r^- y_n(kr)\right) \cdot \left(a_\Theta^+ P_n^m(cos(\Theta)) + a_\Theta^- Q_n^m(cos(\Theta))\right) \\ &\cdot \left(a_\Phi^+ exp(\text{-}jm\Phi) + a_\Phi^- exp(jm\Phi)\right) \end{aligned} \tag{2.136}$$

with the magnitude of the wave number vector being denoted by

$$k = |\mathbf{k}| = \omega\sqrt{\varepsilon\mu}\ . \tag{2.137}$$

Integer numbers *m* and *n* are constants. Special solutions may be defined by choosing a proper set of coefficients $A = \{a_r^+, a_r^-, a_\Theta^+, a_\Theta^-, a_\Phi^+, a_\Phi^-\}$. Superscripts + and exponentials with negative signs, such as $exp(-jm\Phi)$, indicate waves traveling in the positive direction of that coordinate whereas superscripts - and exponentials with positive signs, such as $exp(jm\Phi)$, indicate waves traveling in the negative direction. In (2.136), we employ the **spherical Bessel functions of the first kind**, $j_n(kr)$, and of the **second kind**, $y_n(kr)$, which are related to the regular Bessel function of order $\upsilon = (n+1/2)$ and the **Neumann function** of order $\upsilon = (n+1/2)$, respectively, by

$$\begin{aligned} j_n(kr) &= \sqrt{\frac{\pi}{2kr}} J_{n+1/2}(kr) \\ y_n(kr) &= \sqrt{\frac{\pi}{2kr}} N_{n+1/2}(kr) \end{aligned} \tag{2.138}$$

Also needed in (2.136) are the **associated Legendre functions** of the **first kind**,

$P_n^m(cos(\Theta))$, and of the **second kind**, $Q_n^m(cos(\Theta))$. All these functions and their related representations are tabulated in various mathematical texts such as [9-13].

Finally, according to (2.123), expansion of the magnetic vector potential $\mathbf{A} = \underline{\Psi}\,\mathbf{u_z}$, in spherical coordinates, with respect to the complex-valued vector potential $\underline{\Psi}$ yields the electromagnetic field components TM to the *z* direction as

$$\underline{H}_r = \frac{1}{r}\frac{\partial \underline{\Psi}}{\partial \Phi}, \quad \underline{H}_\Theta = \frac{1}{r}cot(\Theta)\frac{\partial \underline{\Psi}}{\partial \Phi}, \quad \underline{H}_\Phi = -\frac{1}{r}\left(sin(\Theta)\frac{\partial(r\underline{\Psi})}{\partial r} + \frac{\partial(\underline{\Psi}\,cos(\Theta))}{\partial \Theta}\right) \tag{2.139}$$

and

$$\begin{aligned}
\underline{E}_r &= -j\left(\omega\mu\underline{\Psi}\,cos(\Theta) + \frac{1}{\omega\varepsilon}\frac{\partial}{\partial r}\left(\frac{cos(\Theta)}{r^2}\frac{\partial(r^2\underline{\Psi})}{\partial r} - \frac{1}{r\,sin(\Theta)}\frac{\partial(sin^2(\Theta)\underline{\Psi})}{\partial \Theta}\right)\right)\\
\underline{E}_\Theta &= j\left(\omega\mu\underline{\Psi}\,sin(\Theta) - \frac{1}{\omega\varepsilon\, r}\frac{\partial}{\partial \Theta}\left(\frac{cos(\Theta)}{r^2}\frac{\partial(r^2\underline{\Psi})}{\partial r} - \frac{1}{r\,sin(\Theta)}\frac{\partial(sin^2(\Theta)\underline{\Psi})}{\partial \Theta}\right)\right)\\
\underline{E}_\Phi &= -j\frac{1}{\omega\varepsilon\, r\,sin(\Theta)}\frac{\partial}{\partial \Phi}\left(\frac{cos(\Theta)}{r^2}\frac{\partial(r^2\underline{\Psi})}{\partial r} - \frac{1}{r\,sin(\Theta)}\frac{\partial(sin^2(\Theta)\underline{\Psi})}{\partial \Theta}\right)
\end{aligned} \tag{2.140}$$

It is left as an exercise (see Problem 2.3) to calculate, in a dual manner, the **TE** field components by using vector potentials

$$\mathbf{A} = \mathbf{0} \qquad \mathbf{F} = \underline{\Psi}\,\mathbf{u_z} = \underline{\Psi}\left(cos(\Theta)\mathbf{u_r} - sin(\Theta)\mathbf{u_\Theta}\right). \tag{2.141}$$

Note that arbitrary electromagnetic fields can be viewed as superpositions of spherical TE vector components and those determined for TM mode.

▶◀

We shall now see how the components of a time-harmonic electromagnetic field can be described in free space. These components may be computed in spherical coordinates using vector potentials parallel to the *r* direction. Using the orthogonal set of **even** (superscript ^{e}) and **odd** (superscript ^{o}) **tesseral harmonics**

$$\begin{aligned}
T_{m,n}^e(\Theta,\Phi) &= P_n^m(cos(\Theta))\,cos(m\Phi)\\
T_{m,n}^o(\Theta,\Phi) &= P_n^m(cos(\Theta))\,sin(m\Phi)
\end{aligned} \tag{2.142}$$

an arbitrary function can be expanded on the surface of a sphere. $P_n^m(cos(\Theta))$ has already been introduced in Example 2-3 as the associated Legendre function of the first kind. For a constant radius *r* and angles

$0 \le \Theta \le \pi$, $0 \le \Phi \le 2\pi$, only integer numbers m and n produce a finite field. These numbers are conventionally termed "**modes of free space**" [4].

For fields **TE** to spherical coordinate r, a set of vector potentials (with superscript ^{TE}) may be defined by letting

$$\mathbf{A} = \mathbf{0}, \qquad \mathbf{F} = \Psi_r^{\mathrm{TE}} \mathbf{u}_r = (\Psi_r^{\mathrm{TE}}, 0, 0)^T \tag{2.143}$$

and

$$\Psi_r^{\mathrm{TE}} = \begin{Bmatrix} \hat{H}_n^{(1)}(kr) \\ \hat{H}_n^{(2)}(kr) \end{Bmatrix} \begin{Bmatrix} T_{m,n}^{e}(\Theta,\Phi) \\ T_{m,n}^{o}(\Theta,\Phi) \end{Bmatrix} \quad m = 0,1,2,\cdots,n; \quad n = 1,2,\cdots \tag{2.144}$$

where, according to Schelkunoff 's definition [14],

$$\begin{aligned} \hat{H}_n^{(1)}(kr) &= \sqrt{\frac{\pi\, kr}{2}} H_{n+\frac{1}{2}}^{(1)}(kr) = kr\, h_n^{(1)}(kr) \\ \hat{H}_n^{(2)}(kr) &= \sqrt{\frac{\pi\, kr}{2}} H_{n+\frac{1}{2}}^{(2)}(kr) = kr\, h_n^{(2)}(kr) \end{aligned} \tag{2.145}$$

represent the **spherical Hankel functions** of the **first kind** (superscript $^{(1)}$) and of the **second kind** (superscript $^{(2)}$), respectively. We consider functions $\hat{H}_n^{(1)}(kr)$ for waves traveling in the direction of -r, i.e., towards the origin. Waves traveling in the +r direction, i.e., away from the origin, are represented by spherical Hankel functions of the second kind $\hat{H}_n^{(2)}(kr)$. All three components of the electrical field TE to r and in mode m,n are obtained from Ψ_r^{TE} by calculation of

$$\mathbf{E}_{m,n}^{\mathrm{TE}} = -\nabla \times (\Psi_r^{\mathrm{TE}} \mathbf{u}_r). \tag{2.146}$$

Subsequent to the electrical field, the magnetic field vector TE to direction r is computed as

$$\mathbf{H}_{m,n}^{\mathrm{TE}} = j\frac{1}{\omega\mu}\left(\nabla \times \mathbf{E}_{m,n}^{\mathrm{TE}}\right). \tag{2.147}$$

Note that the curl operations in (2.146) and (2.147) must be performed with respect to spherical coordinates. Using (2.144), (2.146), and (2.147) we find the radially directed **wave impedances** for outward-traveling (index +r)

TE modes as

$$\underline{Z}_{+r}^{\mathrm{TE}} = \frac{E_{\Theta}^{+}}{H_{\Phi}^{+}} = -\frac{E_{\Phi}^{+}}{H_{\Theta}^{+}} = -j\eta \frac{\hat{H}_n^{(2)}(kr)}{\hat{H}_n^{(2)}{}'(kr)} \tag{2.148}$$

where the prime stands for the partial derivative ($\partial.../\partial r$) and $\eta = \sqrt{\mu/\varepsilon}$ is the medium's **intrinsic impedance**. Derivatives of cylinder functions of order n are calculated using the recurrence relationships

$$C_n'(z) = C_{n-1}(z) - \frac{n}{z}C_n(z) = -C_{n+1}(z) + \frac{n}{z}C_n(z) \tag{2.149}$$

where **cylinder functions** $C_n(z)$ defined in Eqs. (2.108) and (2.109) denote one of the Bessel functions of order n $J_n(z)$ or $Y_n(z)$, or one of the Hankel functions of order n, $H_n^{(1)}(z)$ or $H_n^{(2)}(z)$.

Similarly, if we need the wave impedance of an inward-traveling (index $-r$) wave in TE mode, this quantity is given by

$$\underline{Z}_{-r}^{\mathrm{TE}} = -\frac{E_{\Theta}^{-}}{H_{\Phi}^{-}} = \frac{E_{\Phi}^{-}}{H_{\Theta}^{-}} = j\eta \frac{\hat{H}_n^{(1)}(kr)}{\hat{H}_n^{(1)}{}'(kr)}. \tag{2.150}$$

By simply interchanging the roles of vector potentials **A** and **F**, we evaluate the field **TM** to radius r. Hence, let us take

$$\mathbf{A} = \Psi_r^{\mathrm{TM}}\mathbf{u}_r = (\Psi_r^{\mathrm{TM}},\ 0,\ 0)^T\mathbf{0}, \qquad \mathbf{F} = \mathbf{0} \tag{2.151}$$

with a set of solutions given by

$$\Psi_r^{\mathrm{TM}} = \begin{Bmatrix} \hat{H}_n^{(1)}(kr) \\ \hat{H}_n^{(2)}(kr) \end{Bmatrix} \begin{Bmatrix} T_{m,n}^{e}(\Theta,\Phi) \\ T_{m,n}^{o}(\Theta,\Phi) \end{Bmatrix} \quad m = 0,1,2,\cdots,n; \quad n = 1,2,\cdots \tag{2.152}$$

Then, we first calculate the magnetic field vector

$$\mathbf{H}_{m,n}^{\mathrm{TM}} = \nabla \times (\Psi_r^{\mathrm{TM}}\mathbf{u}_r) \tag{2.153}$$

and continue to find the electrical field as

$$\mathbf{E}_{m,n}^{\mathrm{TE}} = -j\frac{1}{\omega\varepsilon}\left(\nabla\times\mathbf{H}_{m,n}^{\mathrm{TE}}\right). \tag{2.154}$$

Sets of spherical modes TM to radius r have their own radially directed wave impedances. From Eqs. (2.152), (2.153), and (2.154) it is apparent that

$$\underline{Z}_{+r}^{\mathrm{TM}} = \frac{E_{\Theta}^{+}}{H_{\Phi}^{+}} = -\frac{E_{\Phi}^{+}}{H_{\Theta}^{+}} = j\eta\frac{\hat{H}_n^{(2)\prime}(kr)}{\hat{H}_n^{(2)}(kr)} \tag{2.155}$$

for the outward-traveling wave and

$$\underline{Z}_{-r}^{\mathrm{TM}} = -\frac{E_{\Theta}^{-}}{H_{\Phi}^{-}} = \frac{E_{\Phi}^{-}}{H_{\Theta}^{-}} = -j\eta\frac{\hat{H}_n^{(1)\prime}(kr)}{\hat{H}_n^{(1)}(kr)} \tag{2.156}$$

for the inward-traveling wave. It is interesting to note that radially directed wave impedances are independent of order m of the tesseral harmonics. For a given radius r and a mode number n, there is a so-called **point of gradual cutoff** [4] defined by

$$kr = n \tag{2.157}$$

where the wave impedance changes from being predominantly reactive ($kr < n$) into being predominantly resistive ($kr > n$). By using L. J. Chu's **partial fraction expansion** [15 - 18], a radially directed wave impedance can be represented in the form of a (*Cauer*-) **canonic ladder network.** In a medium with intrinsic wave impedance η (i.e., with magnetic permeability μ and permittivity ε), for a fixed radius r and modal order n, we may write the **TM** wave impedance $\underline{Z}_{+r}^{\mathrm{TM}}$ of an outward-traveling wave in the form of a continued fraction as

$$\left.\underline{Z}_{+r}^{\mathrm{TM}}=\eta\{-j\frac{n}{kr}+\cfrac{1}{j\frac{(1-2n)}{kr}+\cfrac{1}{j\frac{(3-2n)}{kr}+\cfrac{\ddots}{+\cfrac{1}{-j\frac{3}{kr}+\cfrac{1}{-j\frac{1}{kr}+1}}}}}\right\}\quad(2.158)$$

In a dual manner, the spherical wave impedances **TE** to radius r can be expressed in the form of

$$\left.\underline{Z}_{+r}^{\mathrm{TE}}=\eta\{\cfrac{1}{-j\frac{1}{kr}+\cfrac{1}{-j\frac{(2n-1)}{kr}+\cfrac{1}{-j\frac{(2n-3)}{kr}\cfrac{\ddots}{+\cfrac{1}{-j\frac{3}{kr}+\cfrac{1}{-j\frac{1}{kr}+1}}}}}}\right\}\quad(2.159)$$

The circuit diagrams of equivalent highpass filters with resistive loads $R = \eta$ are depicted in Figure 2-5. Each increasing modal term adds another capacitor and inductor to the circuit. For $kr > n$ almost all of the spherical wave's average electric and magnetic energy is "stored" in the capacitors and inductors. Conversely, if $kr < n$, we observe that larger parts of the average energies are transmitted and absorbed by the load resistor R. The point of cutoff, where $kr = n$, represents the limit between these two physical states. An asymptotic analysis can be used to explain the behavior of the equivalent networks displayed in Figure 2-5 (a) and (b). Both circuits are highpass filters and hence $\underline{Z}_{+r}^{\mathrm{TM}}$ and $\underline{Z}_{+r}^{\mathrm{TE}}$ asymptotically approach the value of load impedance $R = \eta$, as frequency $f \to \infty$. On the other side of the spectrum, i.e., for very low frequencies, we observe that $\underline{Z}_{+r}^{\mathrm{TM}} \to \dfrac{n}{j\omega C}$ and $\underline{Z}_{+r}^{\mathrm{TE}} \to j\omega L/n$ where $C = \varepsilon\, r$ and $L = \mu\, r$. The wave impedances are then simply represented by a capacitor or an inductor, respectively.

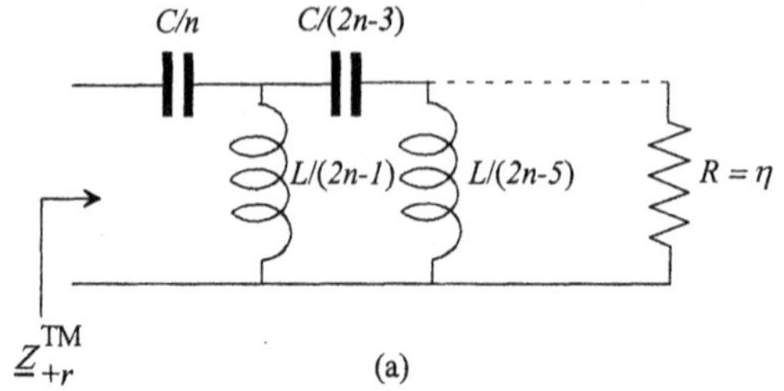

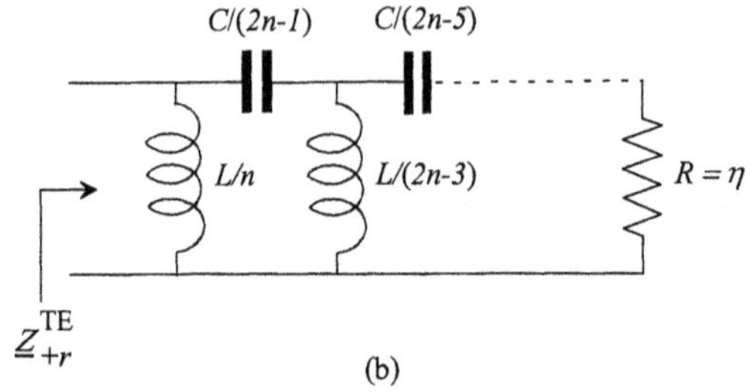

Figure 2-5. Equivalent circuits for (*a*) TM and (*b*) TE spherical wave impedances of modal order n. Reference capacitance $C = \varepsilon r$ and inductance $L = \mu r$.

We saw in Example 2-1 that a homogeneous plane wave in a source-free lossless medium has a single electrical field component (say, E_y) that is oriented transversal to the wave's direction of propagation (e.g, $+z$ in a cartesian system). More generally, it should always be possible, though this is not very convenient, to choose a coordinate system where one axis coincides with the orientation of a particular **E** field vector.

If we want to get rid of the z direction as the arbitrarily chosen direction of wave propagation, we may choose a second cartesian coordinate system (say, with coordinates $\hat{x}, \hat{y}, \hat{z}$) which is rotated relative to the original one with coordinates x, y, z. As an example, the two cartesian systems shown in Figure 2-6 share the same origin, but the axes of the $(\hat{x}, \hat{y}, \hat{z})$ system are rotated.

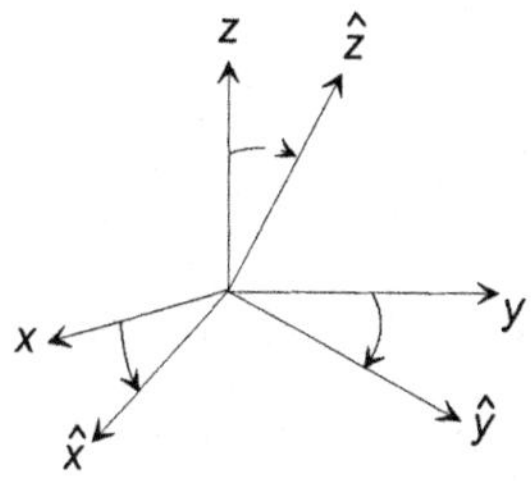

Figure 2-6. Example of two orthogonal cartesian coordinate systems with the ($\hat{x}, \hat{y}, \hat{z}$) system being rotated relative to the fixed (x, y, z) reference system.

By utilization of **direction cosines** $c_{i,j}$ the rotated coordinate system is linked to the (x, y, z) system by an orthogonal matrix in the form of

$$\begin{pmatrix} \hat{x} \\ \hat{y} \\ \hat{z} \end{pmatrix} = \begin{pmatrix} c_{1,1} & c_{1,2} & c_{1,3} \\ c_{2,1} & c_{2,2} & c_{2,3} \\ c_{3,1} & c_{3,1} & c_{3,3} \end{pmatrix} \cdot \begin{pmatrix} x \\ y \\ z \end{pmatrix} \tag{2.160}$$

where, for instance, $c_{1,1} = cos(\angle \mathbf{u}_x, \mathbf{u}_{\hat{x}})$ is the direction cosine of the angle between unit vectors of axes x and $\hat{x}$; next we have $c_{1,2} = cos(\angle \mathbf{u}_x, \mathbf{u}_{\hat{y}})$, $c_{1,3} = cos(\angle \mathbf{u}_x, \mathbf{u}_{\hat{z}})$, and so on. Then a linearly polarized plane wave propagating in the positive $\hat{z}$ direction with amplitude E_+ is described in the ($\hat{x}, \hat{y}, \hat{z}$) system by

$$\mathbf{E} = E_+ \, exp(-jk\hat{z})\mathbf{u}_{\hat{x}}, \tag{2.161}$$

$$\mathbf{H} = \frac{E_+}{\underline{Z}} exp(-jk\hat{z})\mathbf{u}_{\hat{y}} \tag{2.162}$$

where $k = \omega\sqrt{\varepsilon\mu}$. Using **wavenumber vector**

$$\mathbf{k} = k(\, c_{3,1}\mathbf{u}_x + c_{3,2}\mathbf{u}_y + c_{3,3}\mathbf{u}_z \,) = k_x\mathbf{u}_x + k_y\mathbf{u}_y + k_z\mathbf{u}_z \tag{2.163}$$

and (2.160), the electric field vector is converted from the ($\hat{x}, \hat{y}, \hat{z}$) system into the (x, y, z) system by

$$\mathbf{E} = E_+(c_{1,1}\mathbf{u}_x + c_{1,2}\mathbf{u}_y + c_{1,3}\mathbf{u}_z)exp(-jk(c_{3,1}x + c_{3,2}y + c_{3,3}z)) \,. \quad (2.164)$$

For the magnetic field vector, we write

$$\mathbf{H} = \frac{E_+}{\underline{Z}}(c_{2,1}\mathbf{u}_x + c_{2,2}\mathbf{u}_y + c_{2,3}\mathbf{u}_z)exp(-jk(c_{3,1}x + c_{3,2}y + c_{3,3}z)) \,. (2.165)$$

Eqs. (2.164) and (2.165) are somewhat long-winded expressions. Let us therefore try to find a shorter version by letting

$$\mathbf{E}_+ = E_+\mathbf{u}_{\hat{x}} \quad (2.166)$$

and, with respect to (2.164),

$$E_+\mathbf{u}_{\hat{y}} = \mathbf{u}_{\hat{z}} \times \mathbf{E}_+ = \\ = (c_{3,1}\mathbf{u}_x + c_{3,1}\mathbf{u}_y + c_{3,2}\mathbf{u}_z) \times \mathbf{E}_+ = \frac{1}{k}(\mathbf{k} \times \mathbf{E}_+). \quad (2.167)$$

Making use of a position vector $\mathbf{p}$ in the (x, y, z) system, defined by

$$\mathbf{p} = x\mathbf{u}_x + y\mathbf{u}_y + z\mathbf{u}_z \,, \quad (2.168)$$

and the scalar product $\mathbf{k} \cdot \mathbf{p}$, we may rewrite the time-harmonic electromagnetic field of a plane homogeneous wave in a more compact form as

$$\mathbf{E}(\mathbf{p}, t) = \mathbf{E}_+ \, Re\{exp(j(\omega t - \mathbf{k} \cdot \mathbf{p}))\} = \mathbf{E}_+ \, cos(\omega t - \mathbf{k} \cdot \mathbf{p}) \quad (2.169)$$

and

$$\mathbf{H}(\mathbf{p}, t) = \frac{1}{\underline{Z}k}(\mathbf{k} \times \mathbf{E}_+) Re\{exp(j(\omega t - \mathbf{k} \cdot \mathbf{p}))\} = \\ = \frac{1}{\mu\omega}(\mathbf{k} \times \mathbf{E}_+) cos(\omega t - \mathbf{k} \cdot \mathbf{p}) \quad (2.170)$$

We observe that the magnetic field vector has the same space-time dependence as the electric field vector. Note that all dot products $\mathbf{k} \cdot \mathbf{p} = const.$ define planes, and the associated electromagnetic fields represent plane waves. Furthermore, in terms of wavenumber vector $\mathbf{k}$, the

Maxwell equations simplify to

$$\mathbf{H}(\mathbf{p},t) = \frac{1}{\omega\mu}(\mathbf{k}\times\mathbf{E}(\mathbf{p},t)), \tag{2.171}$$

$$\mathbf{E}(\mathbf{p},t) = -\frac{1}{\omega\varepsilon}(\mathbf{k}\times\mathbf{H}(\mathbf{p},t)), \tag{2.172}$$

$$\mathbf{k}\cdot\mathbf{E}(\mathbf{p},t) = 0, \tag{2.173}$$

$$\mathbf{k}\cdot\mathbf{H}(\mathbf{p},t) = 0. \tag{2.174}$$

In other words, all of the three vectors **E**, **H**, and **k** are mutually perpendicular to each other.

The direction of vector **E** is termed **polarization**. The polarization of a wave is conventionally defined by the time variation of the tip of vector **E** at a fixed point in space [19]. Depending on the tip motion over time t, there are various kinds of polarizations. They are categorized as

- *linear* if, at a fixed point on the axis of the wave's propagation, the tip of the **E** field vector traces out a straight line,
- *left-hand* or *right-hand circular* if the tip moves counter-clockwise or clockwise, respectively, on a circle,
- *left-hand* or *right-hand elliptical* if the tip moves counter-clockwise or clockwise, respectively, on an ellipse.

To study polarization effects of time-harmonic plane waves in homogeneous media, we need mathematical models for both the **E** vector field and various forms of tilted ellipses — a special area of the geometry of conic sections. We start by denoting the electric field vector of a time-harmonic plane wave with radian frequency ω and traveling along the positive z axis of a cartesian coordinate system in the form of

$$\begin{aligned}\mathbf{E} &= Re\{\underline{E}_x\, exp(j\omega t - kz)\}\mathbf{u}_x + Re\{\underline{E}_y\, exp(j\omega t - kz)\}\mathbf{u}_y \\ &= \hat{E}_x\, cos(\omega t - kz + \varphi_x)\mathbf{u}_x + \hat{E}_y\, cos(\omega t - kz + \varphi_y)\mathbf{u}_y\end{aligned} \tag{2.175}$$

where $\underline{E}_x = Re\{\underline{E}_x\} + j\ Im\{\underline{E}_x)$ and $\underline{E}_y = Re\{\underline{E}_y\} + j\ Im\{\underline{E}_y]$ are complex-valued quantities with phases

$$\varphi_x = tan^{-1}(\frac{Im\{\underline{E}_x\}}{Re\{\underline{E}_x\}}), \qquad \varphi_y = tan^{-1}(\frac{Im\{\underline{E}_y\}}{Re\{\underline{E}_y\}}) \tag{2.176}$$

and amplitudes

$$\begin{aligned} \hat{E}_x &= \sqrt{(Re\{\underline{E}_x\})^2 + (Im\{\underline{E}_x\})^2} \\ \hat{E}_y &= \sqrt{(Re\{\underline{E}_y\})^2 + (Im\{\underline{E}_y\})^2} \end{aligned} \tag{2.177}$$

Calculation of the cross product (see Eq. (2.171)) of wave number vector $\mathbf{k} = (0, 0, k_z)$ and $\mathbf{E} = (\hat{E}_x\, cos(\omega t - kz + \varphi_x),\ \hat{E}_y\, cos(\omega t - kz + \varphi_y), 0)$ yields the magnetic field vector

$$\begin{aligned} \mathbf{H} &= Re\{-\frac{\underline{E}_y}{Z} exp(j\omega t - kz)\}\mathbf{u}_x + Re\{\frac{\underline{E}_x}{Z} exp(j\omega t - kz)\}\mathbf{u}_y \\ &= -\frac{\hat{E}_y}{Z} cos(\omega t - kz + \varphi_x)\mathbf{u}_x + \frac{\hat{E}_x}{Z} cos(\omega t - kz + \varphi_y)\mathbf{u}_y \end{aligned} \tag{2.178}$$

where $k = |\mathbf{k}| = k_z$ is the propagation constant and $Z = \sqrt{\mu / \varepsilon}$ is the wave impedance.

For a complete description of the time-varying direction and the magnitude of vector **E** (or **H**, respectively), we start by observing that the tip of **E**, when observed at a fixed point on the *z* axis, traces out an ellipse. The simplest analytic form for an ellipse is obtained when the two symmetry axes coincide with the coordinate axes *x* and *y*. It is then defined by its two **semi-axes** *a* and *b* as

$$\frac{x^2}{a^2} + \frac{y^2}{b^2} = 1, \quad a > b \tag{2.179}$$

where $2a$ is the **major axis** of the ellipse and $2b$ is its **minor axis**. The distance from the center of the ellipse to either **focus** is $\sqrt{a^2 - b^2}$. The lengths of the **latera recta** (chords perpendicular to the major axis and going through the foci) are given by

$$\ell = \frac{2b^2}{a}, \tag{2.180}$$

and the **eccentricity** of an ellipse is defined as

$$e = \frac{1}{a}\sqrt{a^2 - b^2}\,. \tag{2.181}$$

Since all ellipses with the same eccentricity are similar, the shape of an ellipse depends only on the **axial ratio** AR (= ratio of major to the minor axes) given by

$$AR = \pm\frac{a}{b} \qquad 1 \le AR \le \infty\,. \tag{2.182}$$

Parametric representations of ellipses are of the form

$$\mathbf{p}(t) = cos(t)\mathbf{v}_1 + sin(t)\mathbf{v}_2 \tag{2.183}$$

where vectors $\mathbf{v}_1$ and $\mathbf{v}_2$ are the major and minor semi-axes. Other parametric equations used to display an ellipse include (see, e.g., [12]) the **rotational parametric representation** as a two-dimensional vector

$$\mathbf{p}(t) = \frac{1}{1+t^2}\left(a(1-t^2),\, 2bt\right)^{\mathrm{T}} \tag{2.184}$$

or the representation in polar coordinates given by

$$r = \frac{ab}{\sqrt{a^2\,sin^2(\Theta) + b^2\,cos^2(\Theta)}} \tag{2.185}$$

where range r is expressed as a function of elevation angle Θ.

If the major axis of an ellipse is inclined, relative to the positive x axis, by an angle of α, the slope of the minor axis is $-1/\alpha$. In that case, different from Eq. (2.179), the analytic equation for the ellipse rotated about the origin is given by

$$\frac{x^2}{a^2}+\frac{y^2}{b^2}+\frac{xy}{c^2}=1, \quad a>b,\ c^2 \neq 0. \tag{2.186}$$

MATLAB program "**elliptica.m**" on the CD ROM can be used to study – for given amplitudes (E_x and E_y) and phases (φ_x and φ_y) of the **E** vector – the electric field vector changes with elapsing time t or, equivalently, with distance along the z axis. Required inputs are the two amplitudes $\hat{E}_x$ and $\hat{E}_y$ of the time-harmonic electric field, and the phase angles φ_x and φ_y in degrees. It is assumed that the field propagates in the positive z direction and that the two orthogonal vector components are phase shifted by

$$\Delta\varphi=\varphi_x-\varphi_y. \tag{2.187}$$

With a minor modification, the direction of propagation can be changed from $+z$ to $-z$ by simply inverting in the source code the sign of time vector **t** and time-shifting the origin by the number (P) of full cycles. Another drawing option would be to change the number of full cycles (i.e., $P{\cdot}2\pi$) to be displayed in the 3D diagram on the right. The figure shown on the top left-hand side represents a projection of the **E** field vector tip onto the (x,y) plane, i.e., at z = 0. If we denote the vertices of the tilted ellipse on the top left-hand side by $max\{|\underline{E}_x(t)|\}=\hat{E}_x$ and $max\{|\underline{E}_y(t)|\}=\hat{E}_y$, respectively, then the tilt angle δ_x of the ellipse, relative to the positive x axis, can be calculated to be

$$\delta_x=\frac{1}{2}tan^{-1}\left(\frac{2\hat{E}_x^2\hat{E}_y^2}{\hat{E}_x^2+\hat{E}_y^2}cos(\Delta\varphi)\right). \tag{2.188}$$

The length of the **major semi-axis** is given by

$$a=\frac{1}{2}\sqrt{2\left(\hat{E}_x^2+\hat{E}_y^2+\sqrt{\hat{E}_x^4+\hat{E}_y^4+2\hat{E}_x^2\hat{E}_y^2\,cos(2\Delta\varphi)}\right)}. \tag{2.189}$$

The **minor semi-axis** of the tilted ellipse has a length of

$$b = \frac{1}{2}\sqrt{2\left(\hat{E}_x^2 + \hat{E}_y^2 - \sqrt{\hat{E}_x^4 + \hat{E}_y^4 + 2\hat{E}_x^2\hat{E}_y^2 \cos(2\Delta\varphi)}\right)}\,. \tag{2.190}$$

As an illustrative example, let us consider a right-handed elliptically polarized plane wave propagating (clockwise) in the positive z direction. Let the amplitudes along the x and y axes of the tilted ellipse be chosen as $\hat{E}_x = 1.7$ and $\hat{E}_y = 0.7$. Furthermore, let us choose phases $\varphi_x = 0°$ and $\varphi_y = -30°$, i.e, a phase difference of $\Delta\varphi = \varphi_x - \varphi_y = 30°$.

Invoking "**elliptica.m**" and entering the above parameters yields the three diagrams shown in Figure 2-7. Note that the z axis is normalized to full wavelengths $\lambda = 2\pi/k$, hence labeled z/λ.

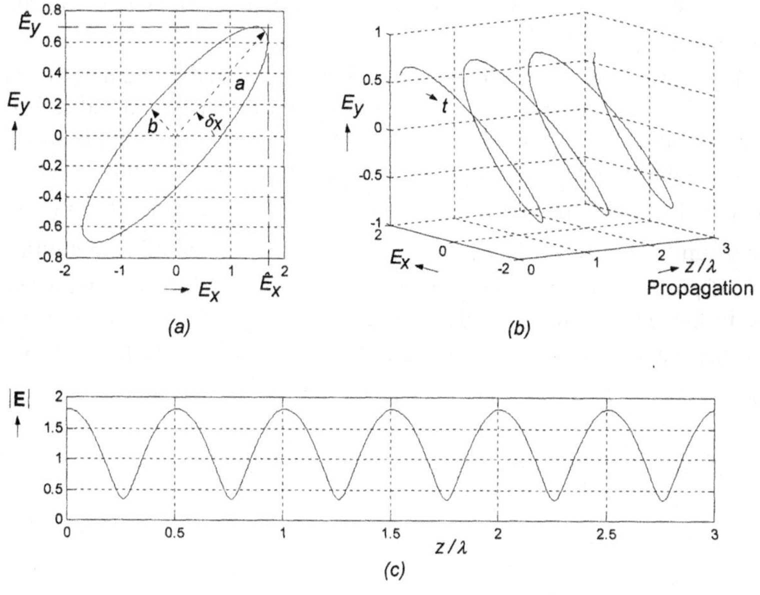

Figure 2-7. Electric field vector **E** of right-hand elliptically polarized wave propagating (clockwise) in positive z direction. Parameters: $\hat{E}_x = 1.7$, $\hat{E}_y = 0.7$, $\Delta\varphi = 30°$. (*a*) E_x and E_y components at $z = 0$ as function of time, tracing out an ellipse tilted by angle δ_x relative to positive x axis. (*b*) Tip of **E** vector as the wave propagates along the positive z axis; distances measured in multiples of wavelength λ. (*c*) Magnitude changes of **E** vector along the z axis.

MATLAB program "**elliptica.m**" also calculates semi-axes a and b, and tilt angle δ_x of the ellipse shown in Figure 2-7*(a)*. For the example under consideration, we obtain a = 1.8085, b = 0.3289, and a tilt angle of δ_x = 20.33°. Because of the right-handed polarization, the sign of axial ratio AR must be negative. Specifically, we have $AR = -a/b = -5.4986$. The sign of AR would be positive had we chosen a left-handed polarization, characterized by a negative time-phase shift of, say, $\Delta\varphi = -30°$.

By simply entering different sets of amplitudes and phases of the two linearly polarized orthogonal electric field vectors, $(\hat{E}_x, \varphi_x)$ and $(\hat{E}_y, \varphi_y)$, the program can serve as a tool to explore various polarization scenarios. If the **E** lines are permanently parallel to the y axis (i.e, $\hat{E}_x = 0$, $\hat{E}_y > 0$), the wave is characterized as **linearly polarized** in the y direction. A more general example of a linearly polarized plane wave is shown in Fig. 2-8. Because of the particular choice of $\hat{E}_x = \hat{E}_y$ and $\Delta\varphi = 0°$, the tip of the wave's electric field vector traces out a straight line inclined by a tilt angle of δ_x = 45°. Note in 2-8*(c)* that the magnitude $|\mathbf{E}|$ periodically vanishes at odd multiples of $\lambda/4$.

If we project the **E** vector of a **circularly polarized wave** onto the (x, y) plane of a standard cartesian system (again with z being the direction of propagation), the ellipse "degenerates" into a circle. An example of a left-hand circularly polarized wave is shown in Figure 2-9. Circular polarization implies that there are two identical electrical field strengths ($\hat{E}_x = \hat{E}_y$), and the magnitude of phase difference $\Delta\varphi$ must be an odd-numbered multiple of 90°. If projected onto the (x, y) plane, the end point of a vector representing the instantaneous electric field now traces out a circle. It is easy to verify that the magnitude $|\mathbf{E}|$ of a circularly polarized wave's electric field vector doesn't change as a function of distance z, i.e.,

$$
\begin{aligned}
|\mathbf{E}| &= \left| Re\{\underline{E}_x\, exp(j\omega t - kz)\}\mathbf{u}_x + Re\{\underline{E}_y\, exp(j\omega t - kz)\}\mathbf{u}_y \right| \\
&= \left| \hat{E}_x\, cos(\omega t - kz + \varphi_x)\mathbf{u}_x + \hat{E}_y\, cos(\omega t - kz + \varphi_y)\mathbf{u}_y \right| \Bigg|_{\substack{\hat{E}_x = \hat{E}_y, \\ |\varphi_x - \varphi_y| = n\frac{\pi}{2};\ n = 1,3,5,\ldots}} \\
&= |\hat{E}_x| = |\hat{E}_y| = const.
\end{aligned}
\tag{2.191}
$$

A circularly polarized wave may hence be defined as a superposition of two orthogonal linearly polarized **E** field components of equal magnitude and a relative phase difference of an odd multiple of $\pi/2$. The reader is encouraged (see Problem 2.5) to check out the necessary conditions for amplitude and phase sets of right- and left-handed plane waves.

It should be clear from the above and from the theory of conic sections that linear and circular polarizations are special cases of elliptical when the axial ratio $\pm a/b$ is zero or $\pm\infty$ ($\rightarrow$ linear polarization), or ± 1 ($\rightarrow$ circular polarization), respectively. Note also that, following Eq. (2.178), the principal time-harmonic behavior of magnetic field vector **H** is almost identical to the one of electric field vector **E**. In a lossless homogeneous medium, the two vectors (i.e., their time-harmonic components) are in-phase (over *time*) and *spatially* orthogonal to each other After calculation of the cross product, it is readily seen that **H** is nothing else but a "replica" of **E**, spatially rotated by 90° and magnitude-scaled by the reciprocal of the medium's wave impedance Z. Vectors **E** and **H** could be displayed in the form of two twisted corkscrews where **E** is always perpendicular to **H** at each point along the axis of propagation. Therefore, loci of the tip of magnetic field vector **H** are usually not shown as separate curves. Also associated with each wave is a power flux that is characterized by **power density vector S** (also known as **Poynting vector**). The instantaneous value of **S** is defined in the form of vector product

$$\mathbf{S} = \mathbf{E} \times \mathbf{H}. \tag{2.192}$$

For time-harmonic waves, we use complex-valued field quantities and the vector of **average power density** given by

$$\overline{\mathbf{S}} = \frac{1}{2} Re\{\mathbf{E} \times \mathbf{H}^*\} \tag{2.193}$$

where the overbar $\overline{(...)}$ stands for time-averaging of all vector components, and the star $(...^*)$ denotes that complex conjugates are to be considered of possibly complex-valued vector components.

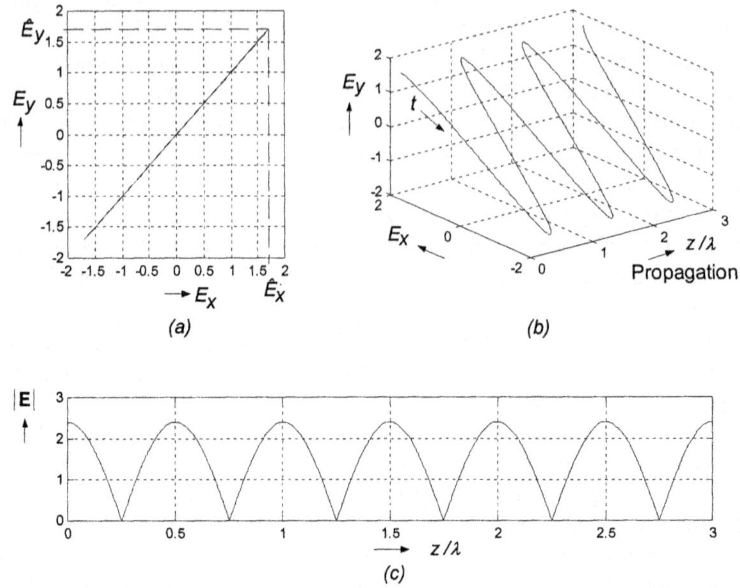

Figure 2-8. Electric field vector **E** of linearly polarized wave propagating along the positive z direction. Parameters: $\hat{E}_x = 1.7$, $\hat{E}_y = 1.7$, $\Delta\varphi = 0°$. (*a*) E_x and E_y components at $z = 0$ as function of time, tracing out a straight line tilted by angle $\delta_x = 45°$ relative to positive x axis. (*b*) Tip of **E** vector as the wave propagates along the positive z axis; distances measured in multiples of wavelength λ. (*c*) Magnitude changes of **E** vector along the z axis.

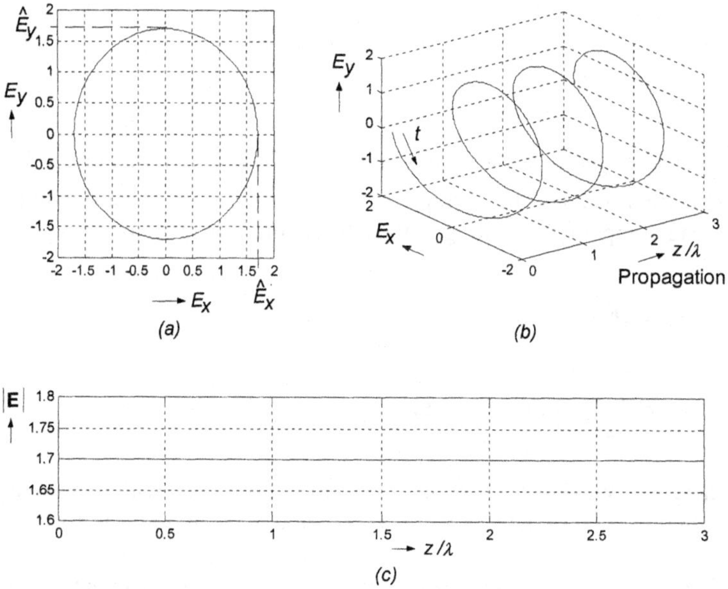

Figure 2-9. Electric field vector **E** of left-hand circularly polarized wave propagating (counter-clockwise) in positive z direction. Parameters: $\hat{E}_x$ = 1.7, $\hat{E}_y$ = 1.7, $\Delta\varphi$ = -90°. (*a*) E_x and E_y components at z = 0 as function of time, tracing out a circle. (*b*) Tip of **E** vector as the wave propagates along the positive z axis; distances measured in multiples of wavelength λ. (*c*) Magnitude changes of **E** vector along the z axis.

Note from the above formulas that the Poynting vector **S** of a *circularly* polarized wave in a lossless, homogeneous medium has a single non-vanishing component parallel to the direction of propagation, that is, unit vector $\mathbf{u}_z$ in our example. By letting $\hat{E}_x = \hat{E}_y = \hat{E}$ and $\Delta\varphi = \varphi_x - \varphi_y = \pm n\pi/2, n = 1,3,5,...,$ we write the electrical field vector of a circularly polarized wave in the form of

$$\mathbf{E} = \begin{pmatrix} \hat{E}\, exp(j(\omega t - kz)) \\ \pm j\,\hat{E}\, exp(j(\omega t - kz)) \\ 0 \end{pmatrix}. \tag{2.194}$$

Since **E** has no z component, the magnetic field vector is readily calculated as

$$\mathbf{H} = \frac{j}{Z}\begin{pmatrix} -\partial \underline{E}_y / \partial z \\ \partial \underline{E}_x / \partial z \\ 0 \end{pmatrix} = \frac{j}{Z}\begin{pmatrix} \pm \hat{E}\,exp(j(\omega t - kz)) \\ -j\hat{E}\,exp(j(\omega t - kz)) \\ 0 \end{pmatrix}. \tag{2.195}$$

Because of the necessary amplitude and phase conditions of the **E** field components involved, the average power density vector $\overline{\mathbf{S}}$ has a single non-vanishing component, pointing in the *z* direction. It is given by

$$\overline{\mathbf{S}} = \frac{1}{2}Re\{\mathbf{E}\times\mathbf{H}^*\} = \frac{1}{2}Re\{\underline{E}_x\underline{H}_y^* - \underline{E}_y\underline{H}_x^*\}\mathbf{u}_z$$
$$= \frac{1}{2}Re\{\frac{\hat{E}}{Z} + \frac{\hat{E}}{Z}\}\mathbf{u}_z = \frac{\hat{E}^2}{Z}\mathbf{u}_z. \tag{2.196}$$

Therefore, we observe that the power flow associated with this wave type is directed parallel to the *z* axis. It is *constant* over both time and space.

Things look much more intricate if two or even a larger number of plane waves with different angular frequencies interact, on the same path, through a lossless, linear and homogeneous medium. The tip of the resulting instantaneous **E** vector of the time-harmonic field is then observed to move along a very strange-looking orbit while traveling through the chosen coordinate system. MATLAB program "**elliptica_mult.m**" is a software tool to explore vector pictures of these multiple-wave scenarios. With no loss of generality, the *z* axis is chosen as the common "railway" for sets of plane waves with variable angular frequencies, polarizations, and possibly opposite directions. The program doesn't set an upper limit to the acceptable number of **E** vectors. Note, however, that the *z* axis is scaled with respect to a unit wavelength or an angular frequency of one, respectively. To give an illustrative example, Figure 2-10 shows the superposition of two elliptically polarized waves, both traveling along the positive *z* direction. The angular frequency (or rotational speed of the vector tip) of one of them is three times higher than that of the other one.

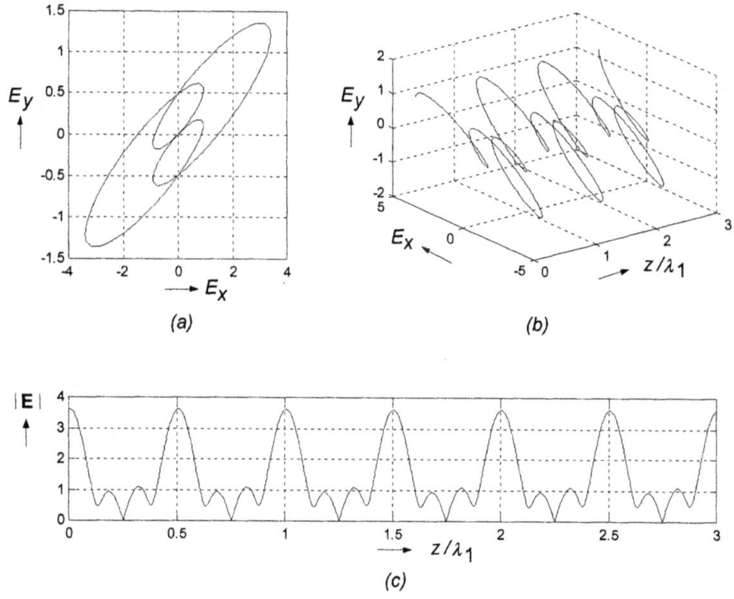

Figure 2-10. Electric field vector **E** of vector sum of two right-hand elliptically polarized waves jointly propagating (clockwise) in positive z direction. Angular frequency of second wave is three times higher than first one. Common parameters of both waves: $\hat{E}_x = 1.7$, $\hat{E}_y = 0.7$, $\Delta\varphi = 30°$. (*a*) Superposition of E_x and E_y components at $z = 0$ as function of time. (*b*) Tip of **E** vector of superimposed waves with z axis normalized by wavelength λ_1 of first wave. (*c*) Magnitude changes of resulting **E** vector along positive z axis.

As shown in Figure 2-11, reversing the direction of propagation of the second wave yields a totally different vector picture. While keeping all other parameters fixed, here, we let the first elliptically polarized wave travel along $+z$ whereas the second one propagates in the negative z direction.

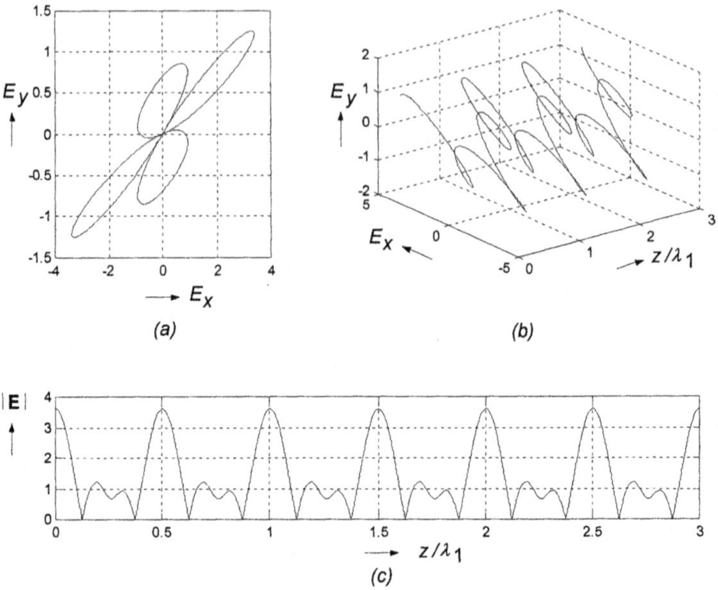

Figure 2-11. Electric field vector **E** of vector sum of two right-hand elliptically polarized waves propagating in opposite directions along z axis. Angular frequency of second wave is three times higher than first one. Common parameters of both waves: $\hat{E}_x = 1.7$, $\hat{E}_y = 0.7$, $\Delta\varphi = 30°$. (*a*) Superposition of E_x and E_y components at $z = 0$ as function of time. (*b*) Tip of **E** vector of superimposed waves with z axis normalized by wavelength λ_1 of first wave. (*c*) Magnitude changes of resulting **E** vector along positive z axis.

2.5 CHAPTER 2 PROBLEMS

2.1 Derive the second vector wave equation (2.29) using $\nabla \times \mathbf{E} = -\mu \dfrac{\partial \mathbf{H}}{\partial t}$ and $\nabla \times \mathbf{H} = \mathbf{J} + \varepsilon \dfrac{\partial \mathbf{E}}{\partial t}$.

2.2 The complex propagation constant is defined as $\underline{\gamma} = \alpha + j\beta$. Show that, according to equations (2.41) and (2.42), the attenuation constant in Np/m is given by $\alpha = \omega\sqrt{\varepsilon\mu}\sqrt{\frac{1}{2}\sqrt{1+\left(\frac{\sigma}{\omega\varepsilon}\right)^2}-1}$ and the phase constant propagation in rad/m is equal to $\beta = \omega\sqrt{\varepsilon\mu}\sqrt{\frac{1}{2}\sqrt{1+\left(\frac{\sigma}{\omega\varepsilon}\right)^2}+1}$.

2.3 Calculate the electric and magnetic field components of a field **TE** to z in a spherical coordinate system using vector potentials

$$\mathbf{A} = \mathbf{0} \qquad \mathbf{F} = \Psi\,\mathbf{u_z} = \Psi\left(cos(\Theta)\mathbf{u_r} - sin(\Theta)\mathbf{u_\Theta}\right).$$

2.4 In free space, we have seen that $k = \omega\sqrt{\varepsilon_0\mu_0}$. For spherical waves in mode n, calculate and plot (using MATLAB) the cutoff radius r_c where $kr = n$ and $n = 1$, 2, 3. Draw the equivalent circuit diagrams of the wave impedances TE and TM to radius r.

2.5 Check out the necessary amplitude and phase conditions for right- and left-handed plane waves. Use MATLAB program "**elliptica.m**" to verify your results.

2.6 Two elliptically polarized plane EM waves of the type shown in Figure 2-7 propagate in *opposite directions* along the z axis of a cartesian coordinate system. Suppose both waves have an angular frequency of ω = 1. Use MATLAB program "**elliptica_mult.m**" to characterize the polarization and the space-time function of the loci of the resulting electrical field vector **E**.

REFERENCES

[1] J. L. Meriam and L. G. Kraige. *Engineering Mechanics — Dynamics*. John Wiley & Sons, Inc., New York, N.Y., 4th edition, 1998.

[2] C. H. Papas. *Theory of Electromagnetic Wave Propagation*. Dover Publications, Inc., Mineola, N.Y., 1988.

[3] C. A. Balanis. *Advanced Engineering Electromagnetics*. John Wiley & Sons, Inc., New York, N.Y., 1989.

[4] R. F. Harrington. *Time-Harmonic Electromagnetic Fields*. IEEE Press and John Wiley & Sons, Inc. New York, N.Y., 2001.

[5] E. Yamashita (Editor). *Analysis Methods for Electromagnetic Wave Problems — Volume Two*. Artech House, Inc., Norwood, MA., 1996.

[6] J. A. Stratton. *Electromagnetic Theory*. McGraw-Hill, New York, N. Y., 1941.

[7] C. R. Scott. *Field Theory of Acousto-Optic Signal Processing Devices*. Artech House, Inc., Norwood, MA., 1992.

[8] Computer Simulation Technology (CST) of America. "Exploring a Three-Dimensional Universe", Microwave Journal, Vol. 44, No. 8, August 2001, pp. 138 - 144. (pdf file available at *www.cst.de/company/news/press/MWjournal_Aug_2001.pdf*).

[9] I. S. Gradstein, I. M. Rhyshik. *Tables of Series, Products, and Integrals*. Vol. 2, Harri Deutsch, Thun, 1981.

[10] M. Abramowitz, I. A. Stegun (eds.). *Handbook of Mathematical Functions with Formulas, Graphs, and Mathematical Tables*. Vol. 55 of National Bureau of Standards Applied Mathematics Series, U. S. Government Printing Office, Washington, D.C., 1964.

[11] M. R. Spiegel and J. M. Liu. *Schaum's Mathematical Handbook of Formulas and Tables*. McGraw-Hill, New York, N.Y., 2nd edition, 1998.

[12] D. Zwillinger. *CRC Standard Mathematical Tables and Formulae*. CRC Press, Boca Raton, FL, 31st edition, 2002.

[13] H. B. Dwight. *Tables of Integrals and Other Mathematical Data*. Prentice Hall, Inc., Englewood Cliffs, N. J., 4th edition, 1961.

[14] S. A. Schelkunoff. *Electromagnetic Waves*. Van Nostrand, Princeton, N.J., 1943.

[15] L. J. Chu. "Physical Limitations of Omnidirectional Antennas," *Journal of Applied Physics*, Vol. 19, December 1948, pp. 1163 - 1175.

[16] J. A. Stratton, P. M. Morse, L. J. Chu, J. D. C. Little, and F. J. Corbato. *Spheroidal Wave Functions*. John Wiley & Sons, Inc., New York, N.Y., 1956.

[17] L. B. Felsen and N. Marcuvitz. *Radiation and Scattering of Waves*. Prentice Hall, Englewood Cliffs, N. J., 1973. (Reprinted by IEEE Press, New York, N.Y., 1996.)

[18] D. M. Grimes and C. A. Grimes. "Minimum Q of Electrically Small Antennas: A Critical Review," *Microwave and Optical Technology Letters,* Vol. 28, No. 3, John Wiley & Sons, February 5, 2001, pp. 172 - 177.

[19] J. A. Kong. *Electromagnetic Wave Theory*. John Wiley & Sons, Inc., New York, N.Y., 2nd edition, 1990.

Chapter 3

ANTENNAS AND RADIATION

3.1 INTRODUCTION

After having dealt with various solutions of Maxwell's equations in particular coordinate systems, let us now proceed with the problems generally encountered in the transmission and reception of single or multiple electromagnetic (EM) waves. Our description of these waves propagating through MIMO channels includes an overview of

- radiation and reception of EM waves by simple antennas,
- different zones and approximations describing fields along the distance from an antenna,
- radiation patterns of antennas, and
- propagation of EM waves through linear media.

While the first three phenomena have to do with the originating sources of EM fields, the fourth category is closely linked to the physical characteristics and the dynamical behavior of the transmission medium itself.

More specifically, our aim is to investigate the radiation patterns of both the transmit and receive antennas. We need explicit formulas describing EM fields distant from a single or multiple known sources. Once these "ideal" fields are sufficiently known, inhomogeneities and perturbation effects can be modeled that are attributable to the non-ideal behavior of the channel. As the physical properties of wireless channels usually change over time, we also need this theoretical background to generate appropriate models using statistical terms or, at least, valid empirical guidelines.

3.2 SIMPLE ANTENNAS AND THEIR EM FIELDS

Let us start our tour through the realms of wireless signal transmission by considering the EM field emitted by a single transmit antenna. The first task is to somehow couple electromagnetic energy from a transmission line (wire, coaxial cable, waveguide, etc.) to free space. Vice versa, at the receiver, the last task in our chain involves conversion of a plane EM wave into one that is guided by a transmission line. Note that, in compliance with the **reciprocity theorem** (see, e.g., [1]), the same rules apply for the EM field properties of transmit and receive antennas.

We could, at least hypothetically and for reference purposes, assume an infinitesimally small antenna with an omnidirectional characteristic. Such an **isotropic radiator** may be conceived as a point source radiating with equal power density in all directions. In geometrical terms, points with equal EM power density are on a sphere with the tiny antenna at its centre. **Directional antennas**, in contrast, radiate or receive EM waves more effectively in some directions than in others. There is an enormous variety of directional antennas, each one with its own geometry and design parameters. Over the past decades, many excellent books have become available on virtually all aspects of practical antenna design. Recommendations include, e.g., [2] - [5]. Theoretical analyses most frequently refer to the classical books by Harrington [6], Balanis [7], Collin [8], to name just a few.

To give a simple example of a, we consider an infinitesimally small electric dipole of **incremental length** $\Delta\ell \to 0$, extending along the z axis from $z = -\Delta\ell/2$ to $z = \Delta\ell/2$. The dipole is surrounded by a homogeneous, lossless, and linear medium. It is excited by a current I flowing in the same direction in each half of the antenna. Thus, the **dipole moment** is $I\Delta\ell$. Time-harmonic fields proportional to $exp(j\omega t)$ are assumed, as usual, and the time dependence of vector potentials and fields will not be explicitly shown below. To get the magnetic vector potential **A**, we write down the Helmholtz equation in the form of

$$\nabla^2 \mathbf{A} + k^2 \mathbf{A} = -\mathbf{J} \tag{3.1}$$

where $k^2 = \omega^2 \varepsilon\mu = (2\pi/\lambda)^2$, and **J** is the vector of electric current density (current/area, A/m^2), due to the radiating source. Vector **A** is oriented parallel to the flow direction of current I, i.e., we denote **A** in cartesian coordinates as $\mathbf{A} = A_z \mathbf{u}_z$. Since the field is generated by a radiating point source, the magnetic vector potential should be spherically symmetric about the origin. It can hence be expressed as a function of radius r in the form of

$$\mathbf{A} = A_z(r)\mathbf{u}_z = A_z\left(cos(\Theta)\mathbf{u}_r - sin(\Theta)\mathbf{u}_\Theta\right). \tag{3.2}$$

With the exception of the origin, i.e., the location of the point source, we may ignore the electric current density **J** on the right-hand side of (3.1) and find that the two solutions of

$$\nabla^2 A_z + k^2 A_z = \frac{1}{r^2}\frac{d}{dr}\left(r^2\frac{dA_z}{dr}\right) + k^2 A_z = 0 \tag{3.3}$$

are proportional to $exp(-jkr)/r$ for an outward-traveling wave and $exp(jkr)/r$ for an inward-traveling wave. To find the proportionality constant, say a, we solve (3.1) for $k \to 0$ and obtain $A_z = I\Delta\ell/(4\pi r)$ which yields $a = I\Delta\ell/(4\pi)$. Then, for an outward-traveling wave, the magnetic vector potential is given by

$$\mathbf{A} = \frac{I\Delta\ell}{4\pi r}exp(-jkr)\mathbf{u}_z = \frac{I\Delta\ell}{4\pi r}exp(-jkr)\left(cos(\Theta)\mathbf{u}_r - sin(\Theta)\mathbf{u}_\Theta\right). \tag{3.4}$$

Substitution of **A** into

$$\mathbf{H} = \nabla \times \mathbf{A} \tag{3.5}$$

and

$$\mathbf{E} = -j\left(\omega\mu\mathbf{A} + \frac{1}{\omega\varepsilon}\nabla(\nabla\cdot\mathbf{A})\right) \tag{3.6}$$

yields a single (*azimuth-oriented*) magnetic field vector component, i.e.,

$$\begin{aligned}\mathbf{H} = H_\Phi\,\mathbf{u}_\Phi &= \frac{I\Delta\ell}{4\pi}exp(-jkr)\left(\frac{1}{r^2} + j\frac{k}{r}\right)sin(\Theta)\mathbf{u}_\Phi \\ &= j\frac{I\Delta\ell}{2\lambda}\frac{sin(\Theta)}{r}exp(-jkr)\left(1 + \frac{1}{jkr}\right)\mathbf{u}_\Phi\end{aligned} \tag{3.7}$$

and an electric field vector **E** with two (*range-* and *elevation-oriented*) non-vanishing components. Vector **E** can be calculated, using (3.6) and the mathematical rules of vector differential calculus, as

$$
\left.
\begin{aligned}
\mathbf{E} &= E_r\mathbf{u}_r + E_\Theta\,\mathbf{u}_\Theta \\
&= \frac{I\Delta\ell}{4\pi}exp(-jkr)\left(2\left(\frac{Z}{r^2} - j\frac{1}{\omega\varepsilon\, r^3}\right)cos(\Theta)\mathbf{u}_r\right. \\
&\qquad \left. + \left(\frac{Z}{r^2} + j\left(\frac{\omega\mu}{r} - \frac{1}{\omega\varepsilon\, r^3}\right)\right)sin(\Theta)\mathbf{u}_\Theta\right) \\
&= jZ\frac{I\Delta\ell}{2\lambda\, r}exp(-jkr)\left(2\,cos(\Theta)\left(\frac{1}{jkr} + \frac{1}{(jkr)^2}\right)\mathbf{u}_r\right. \\
&\qquad \left. + sin(\Theta)\left(1 + \frac{1}{jkr} + \frac{1}{(jkr)^2}\right)\mathbf{u}_\Theta\right)
\end{aligned}
\right\} \tag{3.8}
$$

where $Z = \sqrt{\mu/\varepsilon}$ is the wave impedance of the medium outside the radiating point source.

3.3 FIELD ZONES AND APPROXIMATIONS

Depending on the term kr (or the ratio r/λ, respectively), we distinguish between the following three zones:

- $r/\lambda \ll 1$: *near field (static) zone,*
- $r/\lambda \approx 1$: *intermediate (induction) zone,*
- $r/\lambda \gg 1$: *far field (radiation) zone.*

Close observation of eqs. (3.7) and (3.8) reveals that the fields have totally different properties in these regions. Near the origin, i.e., in the vicinity of the radiating source, the exponential in (3.8) can be replaced by unity, and the field components are quasi-static. Looking at distances far away from the source, the dominating terms in the field equations decay like $1/r$. All terms proportional to $1/r^2$ (or $1/r^3$, respectively) may then be neglected. For sufficiently large values of r/λ, the radial component of **E** vanishes. Hence, the two remaining non-zero components are $E_\Theta(r,\Theta)$ and $H_\Phi(r,\Theta)$. They are mutually orthogonal, in (time) phase relative to each other, and transverse to radius vector $\mathbf{u}_r$. We may, therefore, describe them as radially propagating TEM mode fields. Mathematically, the so-called **far field equations** of the Hertzian electric dipole take on the forms of

$$\left.\begin{aligned}\mathbf{H} &= H_\Phi\,\mathbf{u}_\Phi = jk\frac{I\Delta\ell}{4\pi r}exp(-jkr)\,sin(\Theta)\mathbf{u}_\Phi \\ &= j\frac{I\Delta\ell}{2\lambda}\frac{sin(\Theta)}{r}exp(-jkr)\mathbf{u}_\Phi \\ \mathbf{E} &= E_\Theta\,\mathbf{u}_\Theta = j\frac{I\Delta\ell\,\omega\mu}{4\pi r}exp(-jkr)\,sin(\Theta)\mathbf{u}_\Theta \\ &= jZ\frac{I\Delta\ell}{2\lambda}\frac{sin(\Theta)}{r}exp(-jkr)\mathbf{u}_\Theta = ZH_\Phi\mathbf{u}_\Theta\end{aligned}\right\}\; r \gg \lambda. \tag{3.9}$$

Conversely, we may approximate the **near field equations** by considering in eqs. (3.7) and (3.8) only those terms with the largest decay over range r. The complete set of three non-vanishing field components is obtained as

$$\left.\begin{aligned}\mathbf{H} &= H_\Phi\,\mathbf{u}_\Phi = \frac{I\Delta\ell}{4\pi kr^2}exp(-jkr)\,sin(\Theta)\mathbf{u}_\Phi \\ \mathbf{E} &= E_r\mathbf{u}_r + E_\Theta\,\mathbf{u}_\Theta \\ &= -jZ\frac{I\Delta\ell}{2\pi kr^3}exp(-jkr)\left(cos(\Theta)\mathbf{u}_r + \frac{sin(\Theta)}{2}\mathbf{u}_\Theta\right)\end{aligned}\right\}\; r \ll \lambda. \tag{3.10}$$

Equating the magnitudes of the H_Φ's in the two approximations (3.9) and (3.10), respectively, yields a reasonable limit for the transition from near to far field zones. The two magnetic fields are equally strong (neglecting a 90° phase difference) at the near-to-far field range limit

$$r_{n\leftrightarrow f} = \frac{1}{k} = \frac{\lambda}{2\pi}. \tag{3.11}$$

In the **far field**, the **average power density** (*Poynting*) vector **S** of an electric dipole with its two time-harmonic field components can be calculated as

$$\mathbf{S} = \frac{1}{2} Re\{\underline{E}_\Theta \underline{H}_\Phi^*\}\mathbf{u}_r = \frac{1}{2Z}\left|\underline{E}_\Theta\right|^2 \mathbf{u}_r . \tag{3.12}$$

By integrating **S** over all area elements $d\mathbf{A} = r^2 \sin(\Theta) d\Phi d\Theta \mathbf{u}_r$ of a sphere with constant radius r, we obtain the **average radiated power** as

$$\begin{aligned} \overline{P}_{rad} &= \frac{1}{2Z} \int_{\Theta=0}^{\pi} \int_{\Phi=0}^{2\pi} \left(\frac{Z\left|\underline{I}\right|\Delta\ell}{2\lambda} \frac{\sin(\Theta)}{r} \right)^2 r^2 \sin(\Theta) d\Phi\, d\Theta \\ &= \frac{\pi Z (\left|\underline{I}\right|\Delta\ell)^2}{4\lambda^2} \int_{\Theta=0}^{\pi} \sin^3(\Theta) d\Theta = \frac{\pi}{3} Z \left|\underline{I}\right|^2 \left(\frac{\Delta\ell}{\lambda}\right)^2 \end{aligned} \tag{3.13}$$

Considering the complex-valued impedance $\underline{Z}_a = R_a + jX_a$ at the antenna feedpoint, we have

$$\overline{P}_{rad} = \frac{1}{2} R_a \left|\underline{I}\right|^2 = \frac{\pi}{3} Z \left|\underline{I}\right|^2 \left(\frac{\Delta\ell}{\lambda}\right)^2 \tag{3.14}$$

which yields the **small electric dipole's radiation resistance** in a medium with wave impedance $Z = \sqrt{\mu/\varepsilon}$ as

$$R_a = \frac{2\pi}{3} Z \left(\frac{\Delta\ell}{\lambda}\right)^2 . \tag{3.15}$$

In **free space**, we may substitute the intrinsic impedance of vacuum

$$Z = \eta = \eta_0 = \sqrt{\frac{\mu_0}{\varepsilon_0}} \approx 120\pi \approx 377 \text{ ohms} \tag{3.16}$$

into (3.15) to obtain the following useful approximation of the **radiation resistance**:

$$R_a \approx 80\pi^2 \left(\frac{\Delta\ell}{\lambda}\right)^2 \approx 790 \left(\frac{\Delta\ell}{\lambda}\right)^2 \Omega. \tag{3.17}$$

Note that R_a is independent of the distance from the source, i.e., it doesn't change as a function of range r. Reactance X_a, the imaginary part of $\underline{Z}_a$, can be calculated by integration of the stored evanescent fields.

In a dual manner, the radiation resistance of a **magnetic dipole**, i.e., a small circular current loop with radius r_0, located at the origin, and with magnetic moment

$$M = \pi r_0^2 I \tag{3.18}$$

can be approximated, in **free space** and using far field components (H_Θ, E_Φ), as

$$R_a \approx 320\pi^6 \left(\frac{r_0}{\lambda}\right)^2 \approx 307.6 \left(\frac{r_0}{\lambda}\right)^2 \text{ k}\Omega\,. \tag{3.19}$$

Next, we consider the two most important antenna types shown in Figure 3-1. They are **linear wire antennas** of lengths $\Delta\ell = \lambda/2$ (half-wavelength *dipole* antenna) and $\Delta\ell = \lambda/4$ (quarter-wavelength **monopole antenna** over perfectly conducting surface), respectively.

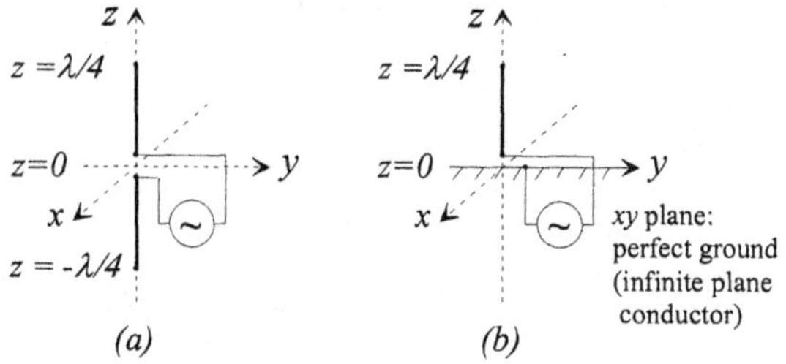

Figure 3-1. Basic elements of simple vertical antennas excited at origin. *(a)* Half-wavelength dipole. *(b)* Quarter-wavelength monopole over perfect ground.

Each of them is excited by a current source I_0 at $z = 0$. Along the upper line segment $0 \le z \le \lambda/4$, the current distributions of both antenna types are sufficiently well approximated as

$$I(z) = I_0\, cos(kz) = I_0\, cos(2\pi \frac{z}{\lambda})\,. \tag{3.20}$$

To determine the EM field, we employ the principle of superposition and integrate over all infinitesimally small electric dipoles along the wire positions. For the purpose of correct integration, we use primed quantities for the *source* coordinates and unprimed *field* coordinates. Therefore, the distance vector $\Delta\mathbf{p}$ between a point on the antenna (at position $\mathbf{p}'$) and a field point (at position $\mathbf{p}$) is defined as

$$\Delta\mathbf{p} = \mathbf{p} - \mathbf{p}'. \tag{3.21}$$

As before, we begin by generating an appropriate z-oriented magnetic vector potential

$$\begin{aligned} \mathbf{A} = A_z\mathbf{u}_z &= \frac{1}{4\pi}\int_{-\lambda/4}^{\lambda/4} \frac{I(z')}{|\Delta\mathbf{p}|} exp(-jk|\Delta\mathbf{p}|)\mathrm{d}z'\,\mathbf{u}_z \\ &= \frac{1}{4\pi}\left(\int_{-\lambda/4}^{\lambda/4} \frac{I(z')}{|\Delta\mathbf{p}|} exp(-jk|\Delta\mathbf{p}|)\mathrm{d}z'\right)\left(cos(\Theta)\mathbf{u}_r - sin(\Theta)\mathbf{u}_\Theta\right) \end{aligned} \tag{3.22}$$

To show details of the geometrical problem, Figure 3-2 considers the two points of interest during integration. Polar source coordinates $\mathbf{p}'$ of an infinitesimally small electric dipole are displayed on the z axis, using *primed* variables. Polar coordinates of EM field vector $\mathbf{p}$ are denoted by *unprimed* variables.

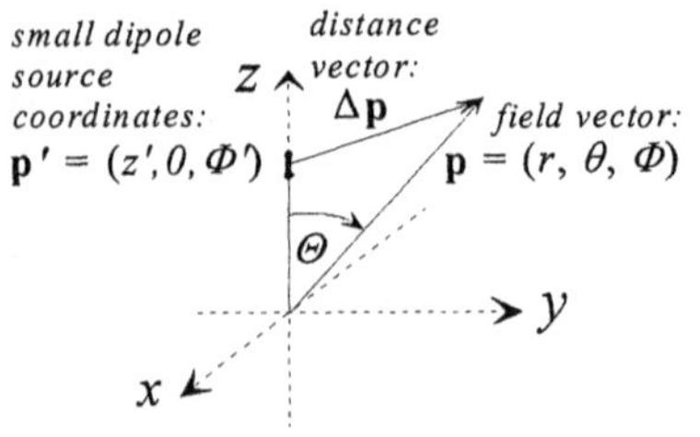

Figure 3-2. Principle of integration over infinitesimally small dipoles along z-oriented thin wire half-wavelength antenna. Spherical source coordinates *primed*; field coordinates *unprimed*.

For an arbitrarily chosen azimuth angle Φ, using polar-to-cartesian conversion formulas

$$\left.\begin{aligned} x &= r\,sin(\Theta)\,cos(\Phi) \\ y &= r\,sin(\Theta)\,sin(\Phi) \\ z &= r\,cos(\Theta) \end{aligned}\right\} \tag{3.23}$$

we may express the cartesian coordinates of **distance vector** $\Delta\mathbf{p}$ as

$$\begin{aligned} \Delta\mathbf{p} &= (x-x')\mathbf{u}_x + (y-y')\mathbf{u}_y + (z-z')\mathbf{u}_z \\ &= r\,sin(\Theta)\,cos(\Phi)\mathbf{u}_x + r\,sin(\Theta)\,sin(\Phi)\mathbf{u}_y + (r\,cos(\Theta)-z')\mathbf{u}_z \end{aligned} \tag{3.24}$$

To sum up, at an EM field point $\mathbf{p}$ of interest, the contributions of all small dipoles, we need the length (magnitude) of $\Delta\mathbf{p}$ which is then to be substituted into the integral of (3.22). It is readily seen that

$$\begin{aligned} |\Delta\mathbf{p}| &= \sqrt{r^2\,sin^2(\Theta)(sin^2(\Phi)+cos^2(\Phi)) + \left(r\,cos(\Theta)-z'\right)^2} \\ &= \sqrt{r^2 + z'^2 - 2rz'\,cos(\Theta)} \end{aligned} \tag{3.25}$$

which, sufficiently far away from the antenna (i.e., in the far field where $(r/z')^2 >> 1$), can be approximated as

$$|\Delta\mathbf{p}| \approx r - z'\,cos(\Theta)\,. \tag{3.26}$$

In the far field, the integral in (3.22) may further be simplified by approximating the denominator of the integrand as $|\Delta\mathbf{p}| \approx r$. By closer inspection of the phase term

$$exp(-jk|\Delta\mathbf{p}|) \approx exp(-j2\pi\frac{r}{\lambda})exp(jkz'\,cos(\Theta)),\ |r| >> |z'| \leq \frac{\lambda}{4} \tag{3.27}$$

we note that a sufficiently accurate far field approximation should at least retain the elevation-dependent exponential. Then, for the current distribution given in (3.20), we end up with a much simpler integral for the magnetic vector potential

$$\mathbf{A} = A_z \mathbf{u}_z \approx \frac{exp(-jkr)}{4\pi\, r} \int_{-\lambda/4}^{\lambda/4} I(z') exp\left(jkz' cos(\Theta)\right) dz'\, \mathbf{u}_z$$

$$= \frac{exp(-jkr)}{4\pi\, r} \left(\int_{-\lambda/4}^{\lambda/4} I_0\, cos(kz') exp\left(jkz' cos(\Theta)\right) dz' \right) \cdot \qquad (3.28)$$

$$\cdot \left(cos(\Theta) \mathbf{u}_r - sin(\Theta) \mathbf{u}_\Theta \right)$$

which, after solving for the definite integral on the right (see Problem 3.2), further reduces to

$$\mathbf{A} = A_z \mathbf{u}_z \approx \frac{exp(-jkr)}{2\pi\, kr} \frac{cos(\frac{\pi}{2} cos(\Theta))}{sin^2(\Theta)} \mathbf{u}_z . \qquad (3.29)$$

Finally, by substituting **A** into (3.5) and (3.6), and retaining only terms $\sim 1/r$, the magnetic far field component is obtained as

$$\mathbf{H} = \left(\nabla \times \mathbf{A}\right) \approx \frac{1}{r} \frac{\partial}{\partial r} \left(r A_\Theta \right) \mathbf{u}_\Phi = \frac{1}{r} \frac{\partial}{\partial r} \left(-r\, sin(\Theta) A_z \right) \mathbf{u}_\Phi$$

$$= j \frac{I_0}{2\pi r} exp(-jkr) \frac{cos(\frac{\pi}{2} cos(\Theta))}{sin(\Theta)} \mathbf{u}_\Phi \qquad (3.30)$$

Knowing that $\mathbf{H} \approx H_\Phi \mathbf{u}_\Phi$ is azimuth-oriented, the electric far field vector must be elevation-oriented with its magnitude being scaled by the medium's wave impedance Z, i.e,

$$\mathbf{E} \approx Z H_\Phi\, \mathbf{u}_\Theta = j \frac{I_0 Z}{2\pi r} exp(-jkr) \frac{cos(\frac{\pi}{2} cos(\Theta))}{sin(\Theta)} \mathbf{u}_\Theta . \qquad (3.31)$$

For a time-harmonic field and using the above far field approximations, the half-wavelength dipole's average power density is given, in the form of a range-oriented average Poynting vector $\overline{\mathbf{S}}$, by

$$\overline{\mathbf{S}} = \frac{1}{2} Re\{\underline{E}_\Theta \mathbf{u}_\Theta \times \underline{H}_\Phi^* \mathbf{u}_\Phi\} = \frac{1}{2} Re\{\underline{E}_\Theta \mathbf{u}_\Theta \times \frac{1}{Z} \underline{E}_\Theta^* \mathbf{u}_\Phi\}$$

$$= \frac{1}{2Z} \left|\underline{E}_\Theta\right|^2 \mathbf{u}_r = \frac{|I_0|^2 Z}{8\pi^2 r^2} \frac{cos^2(\frac{\pi}{2} cos(\Theta))}{sin^2(\Theta)} \mathbf{u}_r \tag{3.32}$$

Integration over all area elements $d\mathbf{A} = r^2 sin(\Theta) d\Phi d\Theta \mathbf{u}_r$ of a sphere with constant radius r then yields the average radiated power as

$$\overline{P}_{rad} = \iint_{\Theta,\Phi} \overline{\mathbf{S}} \cdot d\mathbf{A}$$

$$= \frac{|I_0|^2 Z}{8\pi^2 r^2} \int_{\Theta=0}^{\pi} \int_{\Phi=0}^{2\pi} \left(\frac{cos(\frac{\pi}{2} cos(\Theta))}{sin(\Theta)} \right)^2 r^2 sin(\Theta) d\Phi d\Theta \tag{3.33}$$

$$= \frac{|I_0|^2 Z}{4\pi} \int_{\Theta=0}^{\pi} \frac{cos^2(\frac{\pi}{2} cos(\Theta))}{sin(\Theta)} d\Theta \approx \frac{|I_0|^2 Z}{4\pi} \cdot 1.2188$$

Note that $\overline{\mathbf{S}}$ and $\mathbf{A}$ are vectors, and the dot product yields a scalar quantity $\overline{P}_{rad}$, the (time-) average radiated power.

As there is no straight-forward solution available for the integral on the right of (3.33), one possible way is to make use of the identity

$$\int_{\Theta=0}^{\pi} \frac{cos^2(\frac{\pi}{2} cos(\Theta))}{sin(\Theta)} d\Theta = \frac{1}{2}(\gamma + 2 ln(\pi) - ln(\frac{\pi}{2}) - C_i(2\pi)) \tag{3.34}$$

$$\approx 1.2188$$

where $\gamma \approx 0.57721564...$ is the **Euler's constant** and

$$C_i(x) = \gamma + ln(x) + \int_0^x \frac{cos(\xi) - 1}{\xi} d\xi \tag{3.35}$$

is the cosine integral which can be evaluated by means of program **"cosint.m"** in MATLAB's Symbolic Math Toolbox.

Using (3.33), the radiation resistance of the **half-wavelength dipole** in free space ($Z \approx 120\pi\ \Omega$) is obtained as

$$R_{a,\lambda/2} = \frac{2\overline{P}}{|I_0|^2} \approx \frac{Z}{2\pi} 1.2188 \approx 73.13\,\Omega. \tag{3.36}$$

In a similar manner, the physical properties of a quarter-wavelength monopole (geometry shown in Figure 3-1 (*b*)) can be evaluated. Its field is identical to that of the half-wavelength dipole in the upper half-space ($z \geq 0$). There is no field, however, in the lower half-space ($z < 0$) below the infinite, perfectly conducting ground plane. Assuming the same amplitude I_0 and current distribution in the upper half-space, the monopole's average radiated power is only one half of that of the dipole. Its radiation resistance is, therefore, also reduced by 50 per cent, i.e.,

$$R_{a,\lambda/4} = \frac{1}{2} R_{a,\lambda/2} \approx 36.56\,\Omega. \tag{3.37}$$

3.4 RADIATION PATTERNS

Depending on their **E** and **H** field components, all common antennas have characteristic **radiation patterns**. These patterns typically indicate the antenna's radiation intensities as a function of azimuth (horizontal plane) and elevation (vertical plane), respectively. One possible way to exhibit the angular characteristics of an antenna's far field is to use the **radiation field pattern** which is a 3D plot of

$$RFP(\Theta,\Phi) = \left. \frac{|\mathbf{E}(\Theta,\Phi)|}{max\{|\mathbf{E}(\Theta,\Phi)|\}} \right|_{r=const.} \tag{3.38}$$

The reason why the magnitude of **E** should be normalized and calculated for a fixed range r is that $|\mathbf{E}(\Theta,\Phi)|$ may approach infinity for small values of r. As an example, in Figure 3-3, the characteristic **E** field of a half-wavelength wire dipole is plotted versus both elevation angle Θ and range r, using the far field approximation of (3.31). Note that this approximation of $|\mathbf{E}(r,\Theta)|$ doesn't make sense for small values of r.

For a fixed range r and normalized to the maximum of $|\mathbf{E}|$, we display the *RFP* in a 3D plot over both azimuth and elevation angles (Figure 2-15 *(a)*) or, since the thin wire dipole's *RFP* is independent of azimuth, in the elevation plane as a 2D plot of $|\mathbf{E}(\Theta)|$ (Figure 3-4 *(b)*). Alternatively, we may draw a polar diagram of $|\mathbf{E}(\Theta)|$ (Figure 3-5).

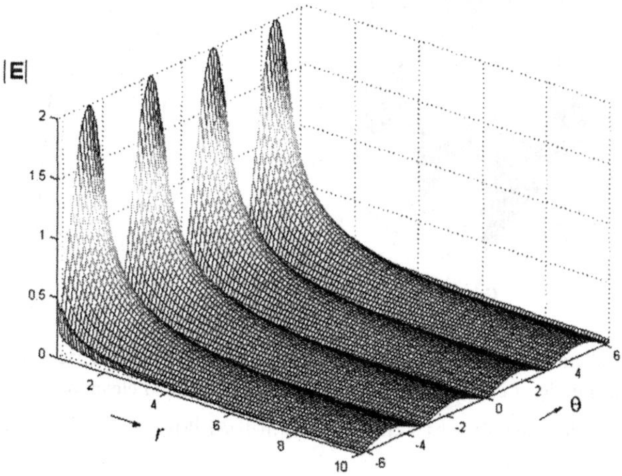

Figure 3-3. Far field approximation of magnitude of Θ-oriented electric field vector. Drawn for half-wavelength dipole (thin wire along z axis, in free space).

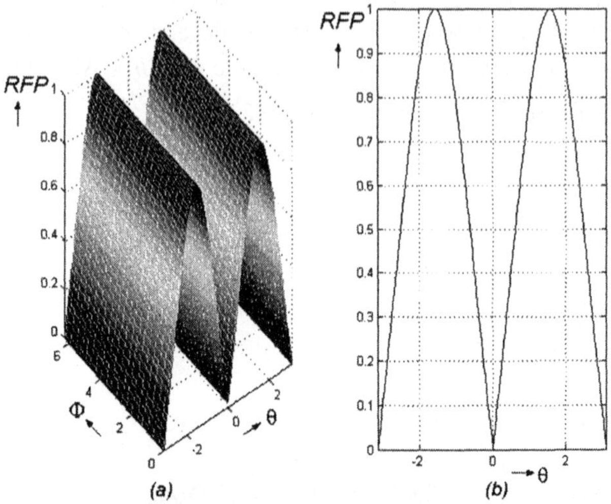

Figure 3-4. Radiation field pattern (*RFP*) of thin wire dipole in free space. *(a)* 3D plot of normalized magnitude of electric field $|\mathbf{E}|$ over both azimuth and elevation angles Φ and Θ, respectively. *(b)* Elevation plane as a 2D plot of normalized $|\mathbf{E}(\Theta)|$.

Amongst RF engineers, it is more customary to show the antenna's **radiation power pattern** (*RPP*), which is the magnitude of the time-average power density vector $\overline{\mathbf{S}}$ (= average Poynting vector) at a fixed range *r*. As the magnitude of $\overline{\mathbf{S}}$ decreases inversely proportional to the square of *r*, we pre-multiply $|\overline{\mathbf{S}}|$ by r^2 and define the far field *RPP* as

$$RPP(\Theta,\Phi) = r^2 \left|\overline{\mathbf{S}}(r,\Theta,\Phi)\right| . \tag{3.39}$$

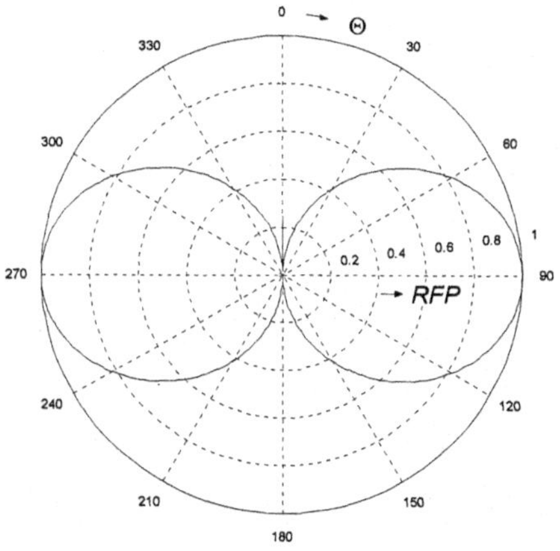

Figure 3-5. Polar plot of radiation field pattern (*RFP*) of half-wavelength wire dipole in free space. Normalized magnitude of electric far field displayed as function of elevation angle Θ.

Using again our half-wavelength dipole example, $\bar{\mathbf{S}}$ doesn't change over azimuth. Thus, if normalized to the maximum of $|\bar{\mathbf{S}}|$ at Θ = 90° and expressed in decibels (superscript *dB*), we obtain

$$RPP_{\lambda/2}^{(dB)}(\Theta) = 10 log_{10}\left(\frac{\frac{|I_0|^2 Z}{8\pi^2 r^2} \frac{cos^2(\frac{\pi}{2}cos(\Theta))}{sin^2(\Theta)}}{\frac{|I_0|^2 Z}{8\pi^2 r^2}} \right) \tag{3.40}$$

$$= 20 log_{10}\left[\frac{cos(\frac{\pi}{2}cos(\Theta))}{sin(\Theta)} \right]$$

A polar plot of (3.40) is displayed in Figure 3-6.

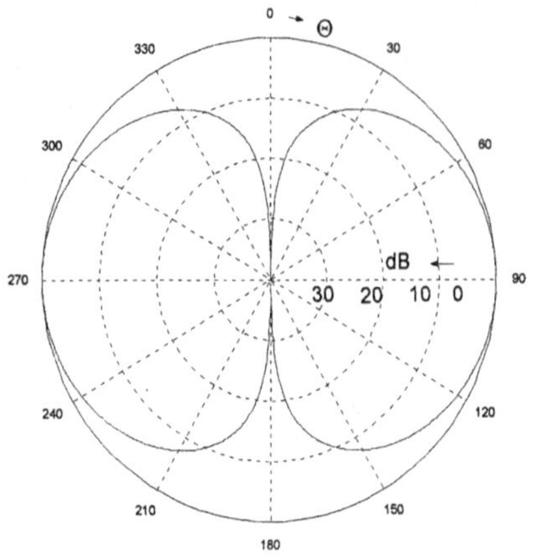

Figure 3-6. Radiation power pattern (*RPP*) of half-wavelength thin wire dipole in free space. Elevation plane view of $RPP_{\lambda/2}^{(dB)}(\Theta)$.

Other radiation power patterns of thin wire dipoles with variable length ratios $\Delta\ell/\lambda$ can be evaluated using MATLAB program "**rpp.m**". Note that this program invokes "**logpolar.m**" which is also available on the CD ROM attached to the book. "logpolar.m" is an extended version of the standard *2D* polar plot program "**polar.m**" included in MATLAB directory "*graph2D*". The user is asked to enter the ratio $\Delta\ell/\lambda$ of interest and *dBrange*, the dynamic range (in dB down the *RPP* maximum). $\Delta\ell$ is the *total* length of the dipole under consideration (not just one arm of it!) and *dBrange* is ten times the logarithm of the dynamic range to be displayed relative to the pattern maximum at 0 dB. Concentric circles are then automatically drawn and labeled in decibels down the maximum. Assuming a sinusoidal current distribution along the *z*-oriented dipole of arbitrary length $\Delta\ell$, (3.40) needs to be modified (see, e.g., [6] or [7]) as follows:

$$RPP_{\Delta\ell}^{(dB)}(\Theta) = 20\,log_{10}\left(\frac{cos\left(\pi(\frac{\Delta\ell}{\lambda})cos(\Theta)\right) - cos\left(\pi(\frac{\Delta\ell}{\lambda})\right)}{sin(\Theta)}\right). \qquad (3.41)$$

Another *RPP* example, now with $\Delta\ell/\lambda = 2$, is shown in Figure 3-7. In case of $\Delta\ell/\lambda = 2$, we find four zeros and four power maxima (0 dB) distributed along the total angular interval of $0 \le \Theta \le 2\pi$.

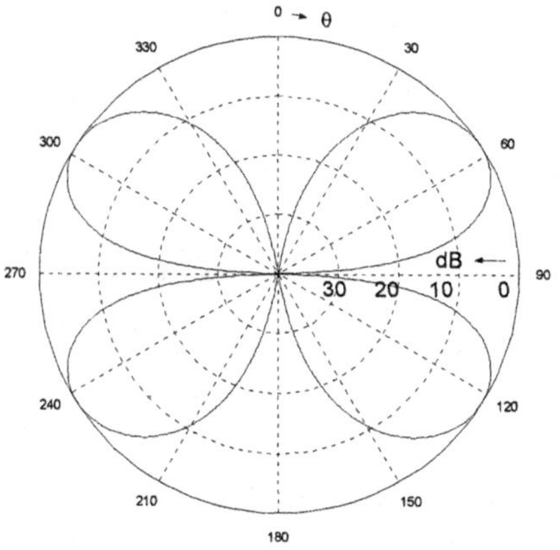

Figure 3-7. Radiation power pattern (*RPP*) of two-wavelengths thin wire dipole $\Delta\ell / \lambda = 2$ in free space. Elevation plane view of $RPP_{2\lambda}^{(dB)}(\Theta)$.

Even more lobes may occur if the relative dipole length $\Delta\ell / \lambda$ goes up. Counting the pattern's total number of lobes, N_L, as a function of $\Delta\ell / \lambda$, we have

$$N_L = \begin{cases} 2L, & \dfrac{\Delta\ell}{\lambda} - L = 0 \\ 2(2L+1), & \dfrac{\Delta\ell}{\lambda} - L > 0 \end{cases} \tag{3.42}$$

where $L = \left\lfloor \dfrac{\Delta\ell}{\lambda} \right\rfloor$ is an integer number obtained by application of the floor operator $\lfloor \ldots \rfloor$, i.e., by rounding $\Delta\ell / \lambda$ towards zero. Note that some of the sidelobes may become very narrow and small if the ratio $\Delta\ell / \lambda$ is close to an integer number. However, they exist and are clearly recognizable if *dBrange* is chosen sufficiently large. To give an example, the fan-shaped radiation pattern shown in Figure 3-8 belongs to an electric dipole with a length ratio of $\Delta\ell / \lambda = 7.1$.

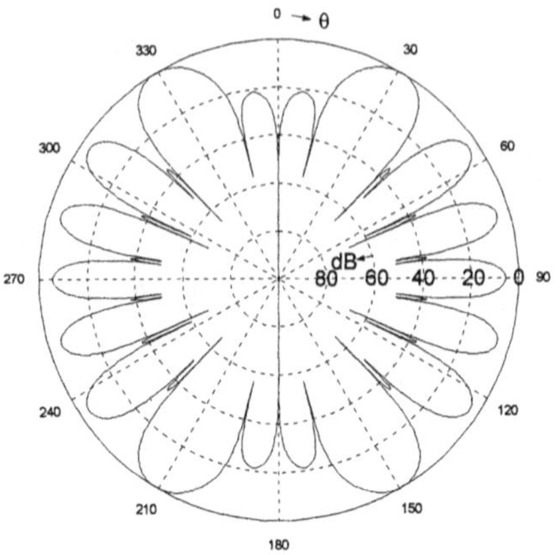

Figure 3-8. Radiation power pattern (*RPP*) of thin wire dipole with length ratio $\Delta\ell/\lambda = 7.1$ in free space. Elevation plane view of $RPP^{(dB)}_{7.1\lambda}(\Theta)$.

Its weak sidelobes next to, e.g, the strong one at Θ = 90° could be overlooked on a linear scale, but are clearly visible on a plot that goes down to 100 dB. The total number of pattern lobes is readily seen to be $N_L = 30$, as determined by (3.42).

The transmit and receive patterns of passive (no amplifiers), reciprocal (i.e., no non-reciprocal elements or materials such as ferrites) antennas are identical.

An antenna's capability to focus its **main beam** in a given direction is usually described by three closely related parameters called **directivity** (D_a), **power gain** (G_a), and **efficiency** (η_a). To calculate, for a given antenna, these three quantities, we recall from solid geometry that there are $4\pi \approx 12.5664$ steradians (sr) in a complete unit sphere [9]. A steradian is, by definition a quantitative aspect attributed to a cone with peak O and extending to infinity. Consider a sphere of radius r centered about point O, as shown in Figure 3-9. The **solid** (or conical) **angle** Ω is defined as the angle subtended at the center of a sphere by an area A on its surface numerically equal to the square of the radius.

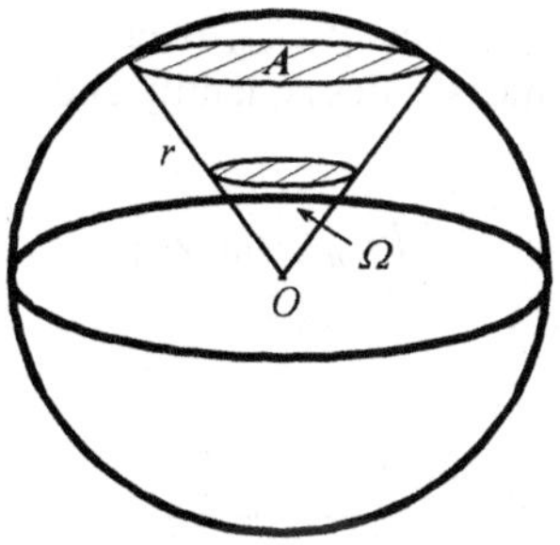

Figure 3-9. Definition of solid angle Ω.

Thus, we have

$$\Omega = \frac{A}{r^2}. \tag{3.43}$$

A unit solid angle (Ω = 1 sr) is obtained if $A = r^2$. It encompasses $1/4\pi \approx 7.9577\,\%$ of the space surrounding a point. Note that the shape of the area A doesn't matter at all. If wish to calculate the solid angle subtended by an arbitrarily shaped surface S, we express Ω in the form of the surface area of a unit sphere covered by the projection of S onto that unit sphere. Using spherical coordinates with differential area $d\mathbf{A} = r^2 \, sin(\Theta) d\Phi d\Theta \mathbf{u}_r$ of a surface patch, we may write Ω in a double integral form as

$$\Omega = \iint_{\Theta,\Phi} \frac{|d\mathbf{A}|}{r^2} = \iint_{\Theta,\Phi} sin(\Theta) d\Phi \, d\Theta \tag{3.44}$$

where, for a given surface S, the integration must be performed over all elevation angles Θ and azimuth angles Φ of that particular surface. Integration over a complete sphere (i.e., $0 \le \Phi \le 2\pi$, $0 \le \Theta \le \pi$) yields Ω = 4π ≈ 12.5664 steradians.

The radiation intensity U_{iso} (in watts per unit solid angle) of an **isotropically radiating source** (one that radiates equally in all directions) is given by the ratio of the average radiated power over the solid angle of the complete sphere, i.e.,

$$U_{iso} = \frac{\overline{P}_{rad}}{4\pi}. \tag{3.45}$$

A *non*isotropic antenna's **maximum directivity** is obtained as the ratio of the maximum radiation intensity U_{max} over the radiation intensity of the isotropic source. It is a dimensionless quantity and can be expressed as

$$D_{a,max} = \frac{U_{max}}{U_{iso}} = \frac{4\pi}{\overline{P}_{rad}} \cdot \left[\frac{RPP_{max} |I_0|^2 Z}{8\pi^2} \right]$$
$$= \frac{|I_0|^2 Z \cdot max\{r^2 |\overline{\mathbf{S}}(\Theta, \Phi)|\}}{2\pi \int\limits_{\Phi=0}^{2\pi} \int\limits_{\Theta=0}^{\pi} |\overline{\mathbf{S}}(\Theta, \Phi)| r^2 \sin(\Theta) d\Theta d\Phi} \tag{3.46}$$

where RPP_{max} is the radiation power pattern taken in the directions (angle Θ, Φ) of its maximum. Of course, directivities $D_a(\Theta,\Phi)$ can also be calculated in any direction of interest, that is, as a function of given angles Θ and Φ, respectively.

An antenna's **maximum power gain** ($G_{a,max}$) is defined as 4π times the ratio of the radiation intensity in the direction of the maximum radiation over the average net input power accepted by the antenna. Mathematically stated, we have the dimensionless quantity

$$G_{a,max} = \frac{4\pi}{\overline{P}_{in}} \cdot \left[\frac{RPP_{max} |I_0|^2 Z}{8\pi^2} \right]$$
$$= \frac{|I_0|^2 Z \cdot max\{r^2 |\overline{\mathbf{S}}(\Theta, \Phi)|\}}{2\pi \overline{P}_{in}} \tag{3.47}$$

where $\overline{P}_{in}$ is the time-average input power at the antenna feed terminals. Note that it is convenient to express both the directivity and the power gain in decibels. Another parameter that is readily derived from $D_{a,max}$ and $G_{a,max}$ is termed **antenna efficiency**, η_a. It is given by the ratio of the maximum power gain ($G_{a,max}$) over the maximum directivity ($D_{a,max}$) or, equivalently, by the ratio of the average radiated power ($\overline{P}_{rad}$) over the average input power ($\overline{P}_{in}$) at the antenna's feed terminals. We observe that $\overline{P}_{rad} \le \overline{P}_{in}$ and $G_{a,max} \le D_{a,max}$. A useful definition of an antenna's total efficiency is thus given by

$$\eta_a = \frac{G_{a,max}}{D_{a,max}} = \frac{\overline{P}_{rad}}{\overline{P}_{in}} = \eta_r \cdot \eta_{cd}, \quad (0 \le \eta_a, \eta_r, \eta_{cd} \le 1) \tag{3.48}$$

where η_a may be split up into a product of two efficiencies. The first one, η_r, is due to reflection losses because of possible mismatch between the antenna (complex impedance: $\underline{Z}_a = R_a + jX_a$) and the connecting cable (transmission line impedance $\underline{Z}_L$). Using the previously defined complex reflection factor, $\underline{r}$, we have

$$\eta_r = 1 - |\underline{r}|^2 = 1 - \left| \frac{\underline{Z}_a - \underline{Z}_L}{\underline{Z}_a + \underline{Z}_L} \right|^2 . \tag{3.49}$$

The second factor, η_{cd}, that potentially decreases the overall efficiency is due to *c*onductive and *d*ielectric losses in the antenna structure. It should be mentioned that η_{cd} is usually slightly less than one and, for almost all practical antennas, difficult to calculate. It is, therefore, more convenient to neglect these small losses (lossless antenna assumption: $\eta_{cd} = 1$) and to consider a small power margin in the overall link budget.

An antenna is in **resonance** if its reactance (= imaginary part of $\underline{Z}_a$) vanishes, i.e., $X_a = 0$.

EXAMPLE 3-1: We wish to calculate the total efficiency, η_a, and the maximum power gain, $G_{a,max}$, of a lossless half-wavelength electric dipole antenna in free space. The antenna is connected to a 50-Ω transmission line.

We start by recalling from (3.40) that the radiation power pattern of the dipole is given by

$$RPP_{\lambda/2}(\Theta) = \frac{cos^2(\frac{\pi}{2} cos(\Theta))}{sin^2(\Theta)} . \tag{3.50}$$

It has its maximum, RPP_{max}, at an elevation angle of Θ = 90°. Hence, $RPP_{max} = max\{RPP_{\lambda/2}(\Theta)\} = RPP_{\lambda/2}(\pi/2) = 1$. The average radiated power was calculated in (3.33) to be

$$\overline{P}_{rad} = \iint_{\Theta,\Phi} \overline{\mathbf{S}} \cdot d\mathbf{A}$$

$$= \frac{|I_0|^2 Z}{8\pi^2 r^2} \int_{\Theta=0}^{\pi} \int_{\Phi=0}^{2\pi} \left(\frac{cos(\frac{\pi}{2} cos(\Theta))}{sin(\Theta)} \right)^2 r^2 sin(\Theta) d\Phi d\Theta \tag{3.51}$$

$$= \frac{|I_0|^2 Z}{4\pi} \int_{\Theta=0}^{\pi} \frac{cos^2(\frac{\pi}{2} cos(\Theta))}{sin(\Theta)} d\Theta \approx \frac{|I_0|^2 Z}{4\pi} \cdot 1.2188 .$$

The dipole's maximum directivity is, according to (3.46), obtained as

$$D_{a,max} = \frac{4\pi}{\overline{P}_{rad}} \left(RPP_{max} \frac{|I_0|^2 Z}{8\pi^2} \right) \approx \frac{16\pi^2 \cdot |I_0|^2 Z \cdot RPP_{max}}{|I_0|^2 Z \cdot 1.2188 \cdot 8\pi^2} \approx 1.641 , \tag{3.52}$$

which is equivalent to $D_{a,max}^{(dBi)} = 10 \cdot log(D_{a,max}) \approx 2.15$ dBi . The pseudo-unit "**dBi**" refers to the fact that $D_{a,max}$ is normalized with respect to the radiation intensity U_{iso} of an **isotropically radiating source**.

The antenna's (ohmic) radiation resistance

$$R_{a,\lambda/2} = \frac{2\overline{P}_{rad}}{|I_0|^2} \approx \frac{Z}{2\pi} \cdot 1.2188 \approx 73.13\,\Omega \tag{3.53}$$

is different from the specified transmission line impedance $Z_L = 50\,\Omega$. Therefore, due to a (power) reflection factor $\underline{r}$, the total antenna efficiency is given by

$$\eta_a = \eta_r = 1 - |\underline{r}|^2 = 1 - \left| \frac{R_{a,\lambda/2} - Z_L}{R_{a,\lambda/2} + Z_L} \right|^2 \approx 0.965 . \tag{3.54}$$

Conductive and dielectric losses (index *cd*) of the antenna are neglected, i.e., we assume $\eta_{cd} = 1$. Finally, we find the maximum power gain of the half-wavelength dipole connected to a 50-Ω transmission line to be

$$G_{a,max} = \eta_a D_{a.max} \approx 1.584 , \tag{3.55}$$

which is equivalent to $G_{a,max}^{(dBi)} \approx 2$ dBi . We note that the mismatch between the radiation resistance of the antenna and the line impedance of the feeding cable causes a small power loss of approximately 0.15 dB. Choice of a 75-Ω transmission line would significantly reduce the reflection coefficient and thus yield almost zero power loss. Actually, we have also ignored the reactance (= imaginary part) of the antenna's impedance at the input terminals.

Although the input reactance of a half-wavelength dipole amounts to $j42.5$ ohms, it can be reduced to zero by making the antenna length slightly shorter than $\lambda/2$. For a detailed discussion and analysis of input impedances of dipoles, the reader is referred to Chapter 7 of [7].

►◄

The sum of an antenna's maximum power gain (in dBi) and the transmit power (in dBm, i.e., referenced to 1 mW) is usually referred to in link budgets as **effective isotropic radiated power**, or ***EIRP***. We may thus express *EIRP* (in dBm) as

$$EIRP = 10\,log(G_{a,max}) + 10\,log(\overline{P}_{in}/1\,\text{mW}) \tag{3.56}$$

where $\overline{P}_{in}$ is the time-average input power at the antenna feed terminals. To give an example, in Europe *EIRP* is generally limited to 20 dBm (or 100 mW) for wireless local area networks (WLAN's). If the maximum antenna gain is 2 dBi, the input power at the antenna feed terminal should then not exceed 18 dBm (or 63.1 mW).

If a lossless antenna is used to receive incoming waves with incident power density $S_{in} = |\mathbf{S}_{in}|$, we may measure the received average power (= $\overline{P}_r$) at the load to which it is connected. Using these two quantities, the ratio

$$A_{eff} = \frac{\overline{P}_r}{S_{in}} \tag{3.57}$$

is defined as the **effective aperture** (or **area**) of the antenna. A_{eff} is not to be confused with the physical aperture. It may sometimes become much larger than the physical area. A good example is a thin wire dipole which, in terms of its capability to intercept incoming waves, "appears" much larger than its physical size might suggest. The effective aperture of an antenna with known maximum directivity $D_{a,max}$ and total antenna efficiency η_a can be shown to be

$$A_{eff} = \frac{\lambda^2}{4\pi}\eta_a D_{a,max} = \frac{\lambda^2}{4\pi}G_{a,max}\,. \tag{3.58}$$

Looking at the half-wave dipole of example 3-1, we can easily calculate the antenna's effective aperture as a function of wavelength λ in the form of $A_{eff} \approx (0.965\cdot 1.641\cdot \lambda^2)/(4\pi) = 0.126\cdot \lambda^2$. That is, A_{eff} increases with the square of wavelength λ, but the aperture decreases with the square of frequency f.

Antenna designers should distinguish carefully between the terms **isotropic** and **omnidirectional**. As the word "*isotropic*" implies, these (hypothetical!) sources radiate equally in all directions, i.e., over the full ranges of both azimuth *and* elevation angles. Associated with them are power densities that are uniformly distributed around large spheres centered on the source. The directivity of an isotropic reference antenna is, therefore, independent of both azimuth angles (Φ) and elevation angles (Θ). A lossless and ideally matched isotropic radiator has unity gain and an effective aperture of

$$A_{eff}^{(iso)} = \frac{\lambda^2}{4\pi} \tag{3.59}$$

for radiation of wavelength λ. In contrast to an isotropic antenna, an *omnidirectional* source radiates uniformly in all azimuth directions, but non-uniformly in the elevation directions. Its radiation power pattern is, in polar coordinates, independent of Φ but not independent of elevation angle Θ. Loosely speaking, if an omnidirectional antenna has less radiation in one direction (e.g., $\Theta = 0°$) it must compensate by emitting more energy along boresight directions. This explains why an omnidirectional antenna focuses radiation in certain elevation directions and, consequently, has a maximum directivity $D_{a,max}$ greater than unity. It should be clear by now that passive antennas cannot generate power but they can, as a result of appropriate pattern design, focus energy into desired cross-sectional coverage areas. By limiting the half-power beamwidth of an antenna's main lobe, we may significantly increase the maximum antenna gain and, simultaneously, the radiation intensity in the focal area.

Suppose we can measure the half-power beamwidths $\Delta\Theta_1$ and $\Delta\Theta_2$, respectively, in the two orthogonal planes. Both quantities may be expressed in radians or, preferably, in degrees. They may or may not be identical, depending on symmetry of the given antenna pattern. Then, the **beam solid angle** $\Delta\Omega_a$ may be approximated as

$$\Delta\Omega_a \cong \Delta\Theta_1 \cdot \Delta\Theta_2 , \tag{3.60}$$

and, using this approximation, the maximum directivity is given by

$$D_{a,max} = 4\pi / \Delta\Omega_a$$

$$\cong \begin{cases} 4\pi / (\Delta\Theta_1 \cdot \Delta\Theta_2), & (\Delta\Theta_1, \Delta\Theta_2 \text{ in radians}) \\ 4\pi(180/\pi)^2 / (\Delta\Theta_1 \cdot \Delta\Theta_2), & (\Delta\Theta_1, \Delta\Theta_2 \text{ in degrees}) \end{cases} \quad (3.61)$$

For **planar antenna arrays** with a single major lobe, Elliott's approximation [10] suggests a factor of 32,400 (instead of $4\pi(180/\pi)^2$ = 41,253) in the numerator on the left of (3.61). For practical antennas, Balanis [7] recommends a good approximation of the maximum antenna gain in the form of

$$G_{a,max} \cong \frac{30{,}000}{\Delta\Theta_1 \cdot \Delta\Theta_2}, \quad (\Delta\Theta_1, \Delta\Theta_2 \text{ in degrees}). \quad (3.62)$$

If the main lobe of the radiation pattern has a symmetrical shape, the half-beamwidth angles are equal and can be calculated for a known $G_{a,max}$ by

$$\Delta\Theta \cong \frac{173.2}{\sqrt{G_{a,max}}}, \quad (\Delta\Theta = \Delta\Theta_1 = \Delta\Theta_2 \text{ in degrees}). \quad (3.63)$$

3.5 PROPAGATION OF EM WAVES THROUGH LINEAR MEDIA

Let us now consider the general scenario usually encountered in fixed (e.g., line-of-sight) or mobile communications. To introduce a few geometrical details, Figure 3-10 shows the positions of the *i*th transmitter (subscript *T,i*) and the *j*th receiver (subscript *R,j*) denoted by position vectors $\mathbf{p}_{T,i}$ and $\mathbf{p}_{R,j}$, respectively.

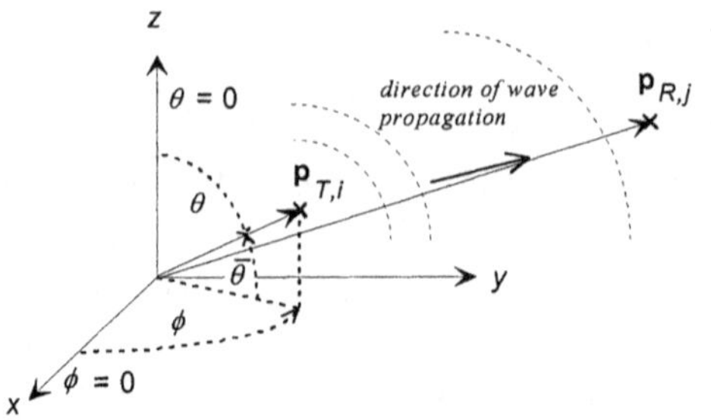

Figure 3-10. Geometry of wave propagation from *i*th transmit antenna (position vector $\mathbf{p}_{T,i}$) to *j*th receive antenna (position vector $\mathbf{p}_{R,j}$) in a polar coordinate reference system.

Assuming a geometrically fixed transmit/receive (Tx/Rx) pair i,j , we may calculate the spatial distance $d_{i,j}$ between the two antennas as the magnitude of the difference vector

$$d_{i,j} = \left|\mathbf{p}_{T,i} - \mathbf{p}_{R,j}\right| . \tag{3.64}$$

To start with the simplest channel model, we choose wave propagation in **free space** and ideal hardware components (cables, filters, etc.), each one having zero insertion loss. These losses may be considered later on by linear loss factors or dB's on a logarithmic scale, respectively. Then, for a given carrier frequency f_0 in Hertz, or wavelength

$$\lambda_0 = \frac{c}{f_0}, \quad (\text{speed of light}: c = 3 \times 10^8 \, \text{m/s}) , \tag{3.65}$$

we wish to calculate the received average power $\overline{P}_{R,j|i}$ at a point described (in polar coordinates) by vector $\mathbf{p}_{R,j}$ and resulting from radiation power $\overline{P}_{T,i}$ emitted by the *i*th transmit antenna. Index i behind the vertical bar in $\overline{P}_{R,j|i}$ denotes the source number. Recalling (3.58), we know that each antenna has an effective aperture A_{eff} and a maximum gain of

$$G_{a,max} = \eta_a D_{a,max} = \frac{4\pi}{\lambda_0^2} A_{eff} = 4\pi (\frac{f_0}{c})^2 A_{eff} . \tag{3.66}$$

The effective isotropic radiated power (*EIRP*) of the *i*th transmitter is given by

$$EIRP_i = \overline{P}_{T,i} \cdot G_{T,i} \tag{3.67}$$

where $G_{T,i}$ is the maximum gain of the ith transmit antenna, as compared to an isotropic antenna with unit gain in all directions. In free space, the average power flux density $\overline{S}_i = |\overline{\mathbf{S}}_i|$ (in W/m²) decreases inversely proportional to the squared distance $d_{i,j}$ from the transmitter. In the far-field, on a sphere with radius $d_{i,j}$ around the ith transmitter, we have

$$\overline{S}_i(d_{i,j}) = \frac{EIRP}{4\pi(d_{i,j})^2} = \frac{\overline{P}_{T,i} \cdot G_{T,i}}{4\pi(d_{i,j})^2} = \frac{|\mathbf{E}_{i,j}|^2}{Z} \quad \text{W/m}^2 \tag{3.68}$$

where Z (= 120π ohms) is the wave impedance of free space and $\mathbf{E}_{i,j}$ is the radiating electric field vector in the far-field region. More precisely, if $\Delta\ell$ is the largest physical dimension of the transmit antenna, we require that the distance between the ith transmitter and the jth receiver exceeds the **Fraunhofer distance**, i.e.,

$$d_{i,j} >> \frac{2(\Delta\ell)^2}{\lambda_0}. \tag{3.69}$$

Multiplication of the average flux density $\overline{S}_i(d_{i,j})$ by the effective aperture

$$A_{eff,j} = \frac{G_{R,j} \cdot \lambda_0^2}{4\pi} \tag{3.70}$$

of the jth receive antenna (max. antenna gain $G_{R,j}$) yields the average received power (in watts) as

$$\begin{aligned} \overline{P}_{R,j} &= \frac{\overline{P}_{T,i} \cdot G_{T,i}}{4\pi(d_{i,j})^2} \cdot A_{eff,j} = \frac{\overline{P}_{T,i} \cdot G_{T,i} \cdot G_{R,j} \cdot \lambda_0^2}{(4\pi)^2 (d_{i,j})^2} \\ &= \overline{P}_{T,i} \cdot G_{T,i} \cdot G_{R,j} \cdot \left(\frac{c}{4\pi f_0 d_{i,j}}\right)^2 \end{aligned} \tag{3.71}$$

Equation (3.71) is widely known, in a two-antenna form omitting overbars and indices i and j, as the **Friis formula**. It relates, for a given carrier wavelength or frequency, respectively, the transmitted power and the received power when two antennas with known gains are located $d_{i,j}$ meters apart from each other. Notice that the Friis model yields sufficiently accurate results in the *far-field region* and in *free space* only. Moreover, in practice, losses due to hardware imperfections need to be considered by a scalar **system loss factor**, say $L_S > 1$, in the denominator of (3.71). It is more convenient to write down the average received power in dBm (i.e., normalized to 1 mW) by taking 10 times the logarithms on both sides of (3.71). Then, we have

$$\overline{P}_{R,j}^{(dBm)} = 10\,log(\frac{\overline{P}_{T,i}}{1\,\mathrm{mW}}) + 10\,log(G_{T,i}) + 10\,log(G_{R,j}) \\ - 20\,log(4\pi) - 20\,log(\frac{f_0 \cdot d_{i,j}}{c}) \tag{3.72}$$

where $-20\,log(4\pi) \approx -21.98\,\mathrm{dB}$. Another quantity of practical importance is the free space **path loss** $\Delta P_{i,j}$ which can be expressed as the ratio of the received power to the transmitted power in the form of

$$\Delta P_{i,j} = \frac{\overline{P}_{R,j}}{\overline{P}_{T,i}} = G_{T,i} \cdot G_{R,j} \cdot (\frac{c}{4\pi\, f_0\, d_{i,j}})^2 \tag{3.73}$$

If expressed in decibels, after changing the signs of the logarithms to obtain positive quantities, the free space path loss can be written as

$$\Delta P_{i,j}^{(dB)} = 20\,log(4\pi) + 20\,log(\frac{f_0\, d_{i,j}}{c}) \\ -10\,log(G_{T,i}) - 10\,log(G_{R,j}) \tag{3.74}$$

Once we know the received power and the (matched!) load resistance R_a at the terminals of the receive antenna, we may calculate the rms voltage V_{rms} available across R_a. The *open circuit* rms voltage of the *unloaded* antenna is

$$V_a = 2\sqrt{\overline{P}_{R,j} \cdot R_a}\,. \tag{3.75}$$

Hence, if the antenna is connected to *load* resistance R_a (i.e. load matched to source resistance), the rms voltage across that load resistance becomes

$$V_{rms} = \frac{V_a}{2} = \sqrt{\overline{P}_{R,j} \cdot R_a} \, . \qquad (3.76)$$

Another quantity of practical interest is the strength of the **electric field** (in V/m) at the terminals of the jth receive antenna. Using equations (3.68) - (3.71), the magnitude of the electric field vector may be calculated as

$$\left| \mathbf{E}_{R,j} \right| = \frac{f_0}{c} \sqrt{\frac{\overline{P}_{R,j} \cdot Z \cdot 4\pi}{G_{R,j}}} = \frac{f_0}{c} \sqrt{\frac{\overline{P}_{R,j} \cdot 480\pi^2}{G_{R,j}}} \quad \text{V/m} \qquad (3.77)$$

where $Z = 120\pi\ \Omega$ is the intrinsic impedance of the transmission medium, i.e., free space.

Let us now go through a typical sequence of Tx/Rx antenna calculations.

EXAMPLE 3-2: Two half-wave dipoles with maximum gains of $G_T^{(dB)} = G_R^{(dB)} = 2$ dB shall be used to transmit a sinusoidal carrier signal of frequency f_0 = 3 GHz and average transmit signal power $\overline{P_T} = 10$ W over a free space distance of d = 5 km. We wish to calculate (*a*) the path loss in dB, (*b*) the received power in Watts and dBm, (*c*) the magnitude of the electric field vector at distance d, and (*d*) the rms voltage available at a load resistance matched to the receive antenna's impedance.

The set of quantities that may be derived from our task description includes:

- wavelength of carrier signal in free space $\lambda = c / f_0 = (3 \cdot 10^8 / 3 \cdot 10^9)\ \text{m} = 10\ \text{cm}$.
- radiation resistance of transmit and receive antenna given by (3.36) as $R_{a,\lambda/2} \approx \frac{Z}{2\pi} \cdot 1.2188 \approx 73.13\ \Omega$.
- Tx/Rx antenna gains are $G_T = G_R = 10^{2/10} \approx 1.585$.

(*a*) Then, using (3.74), index i = 1 for the transmitter, and j = 2 for the receiver, respectively, we find the free space path loss as

$$\Delta P_{1,2}^{(dB)} = 20 log(4\pi) + 20 log(\frac{f_0 d_{1,2}}{c}) - 10 log(G_{T,1}) - 10 log(G_{R,2}) \approx 112\ \text{dB}. \qquad (3.78)$$

(*b*) The average received power in Watts is obtained from (3.71) as

$$\overline{P}_{R,2} = \overline{P}_{T,1} \cdot G_{T,1} \cdot G_{R,2} \cdot (\frac{c}{4\pi f_0 d_{1,2}})^2 \approx 63.6\ \text{pW} \qquad (3.79)$$

which is equivalent to

$$\overline{P}_{R,2}^{(dBm)} = 10 log(\frac{\overline{P}_{R,2}}{1\,\text{mW}}) \approx -72\,\text{dBm}\,. \tag{3.80}$$

(*c*) Using (3.77), we find the magnitude of the electric field vector at the receive antenna as

$$\left|\mathbf{E}_{R,2}\right| = \frac{\pi f_0}{c}\sqrt{\frac{\overline{P}_{R,2}\cdot 480}{G_{R,2}}} \approx 4.36\,\text{mV/m}\,. \tag{3.81}$$

(d) Finally, from (3.76), we find that the rms voltage available at the load resistance R_a (matched to the receive antenna's impedance!) is one half of the open circuit voltage V_a. Thus, under the specified ideal conditions, we should expect a receiver input voltage of

$$V_{rms} = \frac{V_a}{2} = \sqrt{\overline{P}_{R,j}\cdot R_a} \approx 68.2\,\mu\text{V}\,. \tag{3.82}$$

▸◂

3.6 CHAPTER 3 PROBLEMS

3.1 Using an appropriate magnetic vector potential **A**, calculate the EM field equations of a small circular loop of constant current I and with radius r_0, located in the xy plane at the origin, and with magnetic moment $M = \pi r_0^2 I$. (Note the orientation of **A** should be that of the current, i.e., **A** should have a single component in the positive azimuth (Φ) direction.) Show that, in the far field region, the angular (Θ) dependences of the TEM (to r) field components are the same as those of the electric dipole, but the polarization orientations of **E** and **H** are interchanged. Verify the radiation resistance (R_a) formula given in (3.19), based on far field approximations in free space.

3.2 Proof the following identity for the integral on the right of eq. (3.28):

$$\int_{-\lambda/4}^{\lambda/4} I_0\, cos(kz')\, exp\big(jkz' cos(\Theta)\big) dz' = \frac{2}{k}\frac{cos(\frac{\pi}{2}cos(\Theta))}{sin^2(\Theta)}\,.$$

$$Hint\text{: } \int cos(ax)\, exp(bx)\, dx = exp(bx)\frac{b\, cos(ax) + a\, sin(ax)}{a^2 + b^2}\,.$$

3.3 Two half-wave dipoles with maximum gains of $G_T^{(dB)} = G_R^{(dB)} = 1.5\text{ dB}$ shall be used to transmit a sinusoidal carrier signal of frequency f_0 = 1.8 GHz and average transmit signal power $\overline{P_T} = 12\text{ W}$ over a free space distance of d = 500 m. Calculate

(*a*) the path loss in dB,
(*b*) the received power in Watts and dBm,
(*c*) the magnitude of the electric field vector at distance d, and
(*d*) the rms voltage available at a load resistance matched to the receive antenna's impedance.

REFERENCES

[1] S. A. Schelkunoff and H. T. Friis. *Antennas: Theory and Practice*. John Wiley & Sons, Inc., New York, N.Y., 1952.

[2] J. J. Carr. *Practical Antenna Handbook*. 4th edition, McGraw-Hill, New York, N.Y., 2001.

[3] D. Thiel. *Switched Parasitic Antennas for Cellular Communications*. Artech House Books, London, U.K., 2002.

[4] S. N. Makarow. *Antenna and EM Modeling With MATLAB*. John Wiley & Sons, Inc., New York, N.Y., 2002.

[5] G. Kumar and K. P. Ray. *Broadband Microstrip Antennas*. Artech House Books, London, U.K., 2003.

[6] R. F. Harrington. *Time-Harmonic Electromagnetic Fields*. IEEE Press and John Wiley & Sons, Inc. New York, N.Y., 2001.

[7] C. A. Balanis. *Antenna Theory. Analysis and Design*. Harper & Row, Publishers, Inc., New York, N.Y., 1982.

[8] R. E. Collin. *Antennas and Radiowave Propagation*. McGraw-Hill, New York, N.Y., 1985.

[9] The National Institute of Standards and Technology (NIST). *Reference on Constants, Units, and Uncertainty*. Online at *http://physics.nist.gov/cuu/Units/units.html*.

[10] R. S. Elliott. "Beamwidth and Directivity of Large Scanning Arrays," *The Microwave Journal*, January 1964, pp. 74 - 82.

Chapter 4

SIGNAL SPACE CONCEPTS AND ALGORITHMS

4.1 INTRODUCTION

Nowadays, vector space representations of signals offer indispensable tools in modern communications. Geometric interpretations of time-continuous and time-discrete signals are particularly useful in classical signal design and synthesis. Moreover, many applications such as optimum classification and detection of signals, probability-of-error calculations, etc. have greatly benefited from a geometry-based perspective of the technical problem.

4.2 GEOMETRY AND APPLICABLE LAWS

From a geometric viewpoint, we may visualize a real- or complex-valued finite-energy signal $\underline{s}(t)$ as a vector in an orthogonal coordinate system. To get an idea of this fundamental concept, we first recall from the basics of vector analysis that two vectors, say **a** and **b**, are orthogonal (or perpendicular, i.e., $\mathbf{a} \perp \mathbf{b}$) to each other if their dot (or scalar) product is zero. By definition of the dot product, we then have

$$\mathbf{a} \cdot \mathbf{b} = |\mathbf{a}||\mathbf{b}| cos(\angle \mathbf{a}, \mathbf{b}) = 0, \quad 0 \le \angle \mathbf{a}, \mathbf{b} \le \pi \tag{4.1}$$

because of the phasor $\angle \mathbf{a}, \mathbf{b}$, or angle between vectors **a** and **b**, is 90°.

For our purposes, it would be desirable to have a similar definition of orthogonality for pairs of complex-valued functions, or signals of the type $\underline{s}(t) = Re\{\underline{s}(t)\} + j \cdot Im\{\underline{s}(t)\}$, each one with a real and an imaginary part. An appropriate mathematical tool is known as the inner product in a signal vector space. For two ordered pairs of time-continuous complex-valued signals, say $\underline{s}_m(t)$ and $\underline{s}_n(t)$, observed over some finite time interval $[t_1, t_2]$, the **inner product**, will be symbolically denoted in the following by $\langle ... \rangle$. It is defined in integral form as the complex-valued function

$$\langle \underline{s}_m(t), \underline{s}_n(t) \rangle = \int_{t_1}^{t_2} \underline{s}_m(t) \underline{s}_n^*(t)\, dt \tag{4.2}$$

where the asterisk (...*) designates the conjugate of a complex variable. When working with inner products of signals, it is necessary to know and observe a few rules. For any real- or complex-valued scalar weighting factor $\underline{w}$, we note the **scaling laws**

$$\langle \underline{w}\, \underline{s}_m(t), \underline{s}_n(t) \rangle = \underline{w} \langle \underline{s}_m(t), \underline{s}_n(t) \rangle, \tag{4.3}$$

$$\langle \underline{s}_m(t), \underline{w}\, \underline{s}_n(t) \rangle = \underline{w}^* \langle \underline{s}_m(t), \underline{s}_n(t) \rangle. \tag{4.4}$$

By inspection of the definite integral in (4.2), it is readily seen that an interchange of the roles of $\underline{s}_m(t)$ and $\underline{s}_n(t)$, yields the complex conjugate of the inner product. Therefore, the following property of **Hermetian symmetry** holds

$$\langle \underline{s}_m(t), \underline{s}_n(t) \rangle = \langle \underline{s}_n(t), \underline{s}_m(t) \rangle^*. \tag{4.5}$$

Now suppose that signal $\underline{s}_m(t)$ is itself a superposition of a finite number of, say P, signals $\underline{x}_1(t)$, $\underline{x}_2(t)$, ..., $\underline{x}_P(t)$. Similarly, $\underline{s}_n(t)$ may consist of a sum of Q signals $\underline{y}_1(t)$, $\underline{y}_2(t)$, ..., $\underline{y}_Q(t)$. Calculation of the inner product is then accomplished by observing the **distributive laws**

$$\begin{aligned} \langle \underline{s}_m(t), \underline{s}_n(t) \rangle &= \left\langle \sum_{p=1}^{P} \underline{x}_p(t), \underline{s}_n(t) \right\rangle = \sum_{p=1}^{P} \langle \underline{x}_p(t), \underline{s}_n(t) \rangle \\ &= \langle \underline{x}_1(t), \underline{s}_n(t) \rangle + \langle \underline{x}_2(t), \underline{s}_n(t) \rangle + \ldots + \langle \underline{x}_P(t), \underline{s}_n(t) \rangle, \end{aligned} \tag{4.6}$$

$$\begin{aligned} \langle \underline{s}_m(t), \underline{s}_n(t) \rangle &= \left\langle \underline{s}_m(t), \sum_{q=1}^{Q} \underline{y}_q(t) \right\rangle \\ &= \langle \underline{s}_m(t), \underline{y}_1(t) \rangle + \langle \underline{s}_m(t), \underline{y}_2(t) \rangle + \ldots + \langle \underline{s}_m(t), \underline{y}_Q(t) \rangle. \end{aligned} \tag{4.7}$$

Combining the above rules, we may conclude that linear combinations of waveforms can be treated as weighted sums on both sides within in the inner product brackets. Suppose $\underline{s}_m(t)$ is a weighted sum of P complex-valued signals with complex weighting factors $\underline{w}_{m1}$, $\underline{w}_{m2}$, ... $\underline{w}_{mP}$ and $\underline{s}_n(t)$ is a weighted sum of Q complex-valued signals with weights $\underline{w}_{n1}$, $\underline{w}_{n2}$, ..., $\underline{w}_{nQ}$. The inner product of these two **linear combinations** of signals is then calculated as a double sum in the form of

$$\begin{aligned}\langle \underline{s}_m(t), \underline{s}_n(t) \rangle &= \left\langle \sum_{p=1}^{P} \underline{w}_{mp} \underline{x}_p(t), \sum_{q=1}^{Q} \underline{w}_{nq} \underline{y}_q(t) \right\rangle \\ &= \underline{w}_{m1} \underline{w}_{n1}^* \langle \underline{x}_1(t), \underline{y}_1(t) \rangle + \underline{w}_{m1} \underline{w}_{n2}^* \langle \underline{x}_1(t), \underline{y}_2(t) \rangle + \ldots \\ &\quad \ldots + \underline{w}_{mP} \underline{w}_{nQ}^* \langle \underline{x}_P(t), \underline{y}_Q(t) \rangle \\ &= \sum_{p=1}^{P} \sum_{q=1}^{Q} \underline{w}_{mp} \underline{w}_{nq}^* \langle \underline{x}_p(t), \underline{y}_q(t) \rangle \end{aligned} \tag{4.8}$$

Using our knowledge about the calculation of the inner product of two signal segments, we shall now see how these calculations can be used in the context of signal orthogonality.

The energy E_m contained in the mth signal $\underline{s}_m(t)$ measured over a finite time interval $[t_1, t_2]$ is given by the definite integral

$$E_m = \int_{t_1}^{t_2} \underline{s}_m(t) \underline{s}_m^*(t) dt = \int_{t_1}^{t_2} |\underline{s}_m(t)|^2 dt . \tag{4.9}$$

More precisely, E_m is the amount of energy dissipated in a 1 ohm resistance in $\Delta t = t_2 - t_1$ seconds due to a voltage or current signal with waveform $\underline{s}_m(t)$.

4.3 ORTHOGONALITY OF SIGNALS

Any two finite-length observation signals, $\underline{s}_m(t)$ with non-zero energy E_m and $\underline{s}_n(t)$ with non-zero energy E_n, respectively, are said to be **orthogonal** over a limited time interval $[t_1, t_2]$ if

$$\int_{t_1}^{t_2} \underline{s}_m(t)\underline{s}_n^*(t)dt = \begin{cases} E_m = E_n, & m = n \\ 0, & m \neq n \end{cases} \tag{4.10}$$
$$= E_m\delta(m-n) = E_n\delta(m-n)$$

where the Kronecker delta function

$$\delta(m-n) = \begin{cases} 1, & m = n \\ 0, & m \neq n \end{cases} \tag{4.11}$$

is used for the sake of brevity.

We assume further that there exists a set of pairwise mutually orthogonal waveforms $\mathbf{U} = \{\underline{u}_1(t), \underline{u}_2(t), ...\}$ constituting the unit coordinates in our orthogonal coordinate system called **vector signal space**. Application of (4.10) to any two of these waveforms (e.g., those with indices m and n, respectively) yields

$$\int_{t_1}^{t_2} \underline{u}_m(t)\underline{u}_n^*(t)dt = \begin{cases} 1, & m = n \\ 0, & m \neq n \end{cases} \tag{4.12}$$

A natural extension of the concept of orthogonality is the definition of **orthonormal** signals. Two signals are orthonormal if they are orthogonal and, in addition, each one of them has unit norm. In analogy to the length of a vector in Euclidean space (say, $|\mathbf{a}|$ of vector **a**), the natural norm of a signal is defined as the square root of the signal's energy. Using (4.9), the **natural norm** of the mth complex-valued signal $\underline{s}_m(t)$ is given by

$$\|\underline{s}_m(t)\| = \sqrt{\langle \underline{s}_m(t), \underline{s}_m(t)\rangle}$$
$$= \sqrt{E_m} = \sqrt{\int_{t_1}^{t_2} \underline{s}_m(t)\underline{s}_m^*(t)dt} = \sqrt{\int_{t_1}^{t_2} |\underline{s}_m(t)|^2 dt}. \tag{4.13}$$

It is customary to write $\underline{s}_m(t)$ between two double (instead of single) vertical bars. Normalization of a set of signals is always possible by simply dividing each one of the signals by the square root of its energy. If, for instance, energies E_m and E_n of two orthogonal signals $\underline{s}_m(t)$ and $\underline{s}_n(t)$, respectively, are unequal one, we may scale the signals appropriately to make them orthonormal to each other. In terms of their inner product, we write a pair of **orthonormalized** signals as

$$\left\langle \underline{\tilde{s}}_m(t), \underline{\tilde{s}}_n(t) \right\rangle = \int_{t_1}^{t_2} \frac{\underline{s}_m(t)}{\sqrt{E_m}} \cdot \frac{\underline{s}_n^*(t)}{\sqrt{E_n}} dt = \begin{cases} 1, & m = n \\ 0, & m \neq n \end{cases} \qquad (4.14)$$

$$= \delta(m-n)$$

where, to distinguish normalized from unnormalized signals, superscript tilde ($\tilde{\ }$) stands for the property of having unit norm in the given integration time interval $[t_1, t_2]$.

One of the reasons why we often perform the final normalization step is that orthonormal signal sets can be used to expand, over the same finite observation interval $[t_1, t_2]$, an arbitrarily chosen finite-energy waveform, say $\underline{x}(t)$, into a convergent series. Expressed as an infinite sum, we may write $\underline{x}(t)$ by adding up properly weighted versions of all orthonormal signals of that set. Suppose our signal set includes an infinite number of orthonormal complex-valued signals. Let us call the ℓ th one of these orthonormal waveforms $\underline{\tilde{s}}_\ell(t)$, and weight it by a properly chosen complex factor $\underline{w}_\ell$. Then, summation over all weighted signals yields a perfect representation of an arbitrarily chosen $\underline{x}(t)$ in the form of the **generalized Fourier series**

$$\underline{x}(t) = \sum_{\ell=-\infty}^{\infty} \underline{w}_\ell \, \underline{\tilde{s}}_\ell(t) . \qquad (4.15)$$

We call the ℓ th complex weighting factor $\underline{w}_\ell$ a **generalized Fourier coefficient**. Note that these coefficients are meaningful only with respect to a unique and *a-priori* specified set of orthonormal signals. For a known waveform $\underline{x}(t)$, the coefficients may be calculated by multiplication of both sides of (4.15) by the complex conjugate $\underline{\tilde{s}}_\ell^*(t)$ and integration over $[t_1, t_2]$. Doing this and observing the above rules for the calculation of inner products yields the ℓ th generalized Fourier coefficient as

$$\underline{w}_\ell = \int_{t_1}^{t_2} \underline{x}(t) \underline{\tilde{s}}_\ell^*(t) \, dt . \qquad (4.16)$$

Note that all these coefficients may be calculated individually and that the complex conjugate superscript can be omitted if the orthonormal signal set consists of real-valued waveforms. Also, depending on the $\underline{x}(t)$ under consideration and the choice of orthonormal signals, the coefficients may be real or complex scalar quantities. In most practical cases, we have available only a finite number L of orthonormal signals. Clearly, we should then expect a non-zero error between the exact signal $\underline{x}(t)$ and its **approximation** $\underline{x}_a(t)$ given by

$$\underline{x}_a(t) = \sum_{\ell=1}^{L} \underline{w}_\ell \, \underline{\tilde{s}}_\ell(t). \tag{4.17}$$

In vector notation (bold face letters!), we may rewrite (4.17) in an L-dimensional signal vector space as the dot product of **weight vector** (transpose superscript T)

$$\mathbf{w}_L = (\underline{w}_1, \underline{w}_2, \ldots, \underline{w}_L)^T \tag{4.18}$$

multiplied by **orthonormal signal vector**

$$\tilde{\mathbf{s}}_L(t) = (\underline{\tilde{s}}_1(t), \underline{\tilde{s}}_2(t), \ldots, \underline{\tilde{s}}_L(t))^T. \tag{4.19}$$

Thus, the approximated signal $\underline{x}_a(t)$ may be expressed as

$$\underline{x}_a(t) = \mathbf{w}_L^T \cdot \tilde{\mathbf{s}}_L(t). \tag{4.20}$$

The approximation quality can be measured in terms of a suitably defined scalar quantity, preferably a definite integral over the squared error or the square root of that integral. In our previously defined signal vector space, we trace out the exact signal $\underline{x}(t)$, i.e., the one we wish to approximate, and its approximated version $\underline{x}_a(t)$. After having observed these two waveforms between the limits of $[t_1, t_2]$, we consider the complex error signal $\underline{e}(t) = \underline{x}(t) - \underline{x}_a(t)$ and calculate its natural norm, or **error metric**, as

$$\|\underline{e}(t)\| = \sqrt{\langle \underline{x}(t) - \underline{x}_a(t), \underline{x}(t) - \underline{x}_a(t) \rangle} = \sqrt{\int_{t_1}^{t_2} \underline{e}(t)\underline{e}^*(t)dt}. \tag{4.21}$$

Applications of error metrices are quite frequently found in signal processing and communications where minimization of the **error signal energy**, i.e., the squared error metric, is a major goal.

Let us now go through two illustrative examples to learn more about inner products of complex signals, orthogonality, and the process of orthonormalization.

EXAMPLE 4-1: Suppose out of a set **S** of complex signals we take the first two signals

$$\underline{s}_1(t) = \begin{cases} t(1-j) & -1 \le t \le 1 \\ 0 & else \end{cases} \qquad \underline{s}_2(t) = \begin{cases} t(1-j) & -1 \le t \le 0 \\ -t(1-j) & 0 \le t \le 1 \\ 0 & else \end{cases} \tag{4.22}$$

Their real and imaginary parts are depicted in Figure 4-1. Two alternative ways of displaying the given signals are shown in Figure 4-2. We may either draw them in the form of *3D*-plots with time t representing the third axis or show the trajectories (paths in vector space) viewed after a projection onto the two-dimensional complex signal plane, also known as signal constellation diagram.

Are these two complex signals orthogonal to each other over the finite time interval [-1, 1]? If yes, may we classify them as a pair of orthonormal signals?

According to (4.10), to be orthogonal, the two signals' inner product should be zero. It is readily seen that this is the case because

$$\begin{aligned} \left\langle \underline{s}_1(t), \underline{s}_2(t) \right\rangle &= \int_{-1}^{0} [t(1-j)][t(1-j)]^* dt + \int_{0}^{1} [t(1-j)][-t(1-j)]^* dt \\ &= 2\left(\int_{-1}^{0} t^2 dt - \int_{0}^{1} t^2 dt \right) = 0. \end{aligned} \tag{4.23}$$

To be orthonormal, the two signals should have unit energy. The energy values are calculated as

$$E_1 = \left\langle \underline{s}_1(t), \underline{s}_1^*(t) \right\rangle = E_2 = \left\langle \underline{s}_2(t), \underline{s}_2^*(t) \right\rangle = \frac{4}{3} \neq 1. \tag{4.24}$$

Both signals are, in their original forms, *not* orthonormal. We may, however, gain orthonormality by amplitude scaling each one of them by the reciprocal of the square root of its energy. As required by (4.14), we obtain a pair of orthonormal signals in the form of

$$\tilde{\underline{s}}_1(t) = \frac{1}{\sqrt{E_1}} \underline{s}_1(t) = \frac{1}{2}\sqrt{3}\, \underline{s}_1(t), \qquad \tilde{\underline{s}}_2(t) = \frac{1}{\sqrt{E_2}} \underline{s}_2(t) = \frac{1}{2}\sqrt{3}\, \underline{s}_2(t) \tag{4.25}$$

with their inner product being

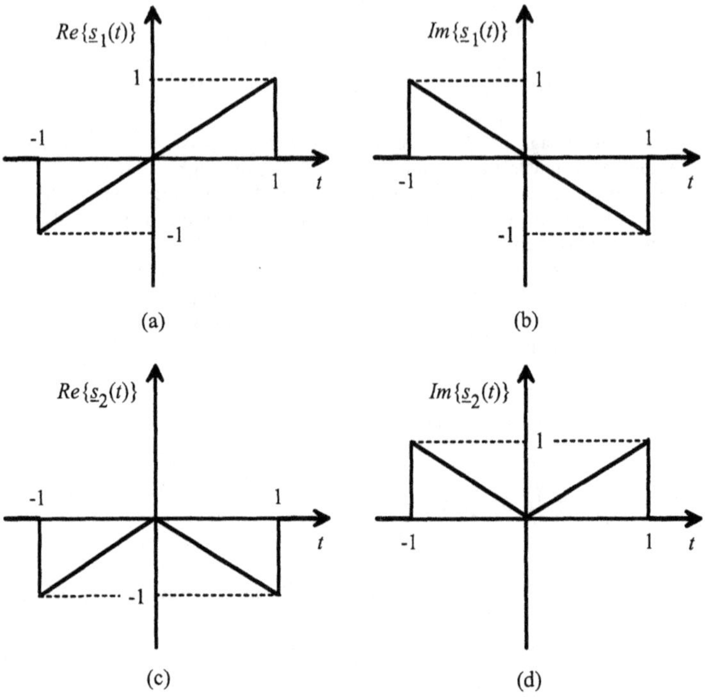

Figure 4-1. Complex signals $\underline{s}_1(t)$ and $\underline{s}_2(t)$ for Example 4-1. Plots of real parts (a), (c), and imaginary parts (b), (d).

$$\left\langle \underline{\tilde{s}}_1(t), \underline{\tilde{s}}_2(t) \right\rangle = \int_{-1}^{1} \frac{\underline{s}_1(t)}{\sqrt{E_1}} \frac{\underline{s}_2^*(t)}{\sqrt{E_2}} dt = \frac{3}{4} \int_{-1}^{1} \underline{s}_1(t) \underline{s}_2^*(t) dt = 0. \tag{4.26}$$

To conclude our example, we may ask the question of whether or not the given signals remain orthogonal to each other if they are observed over a different time window.

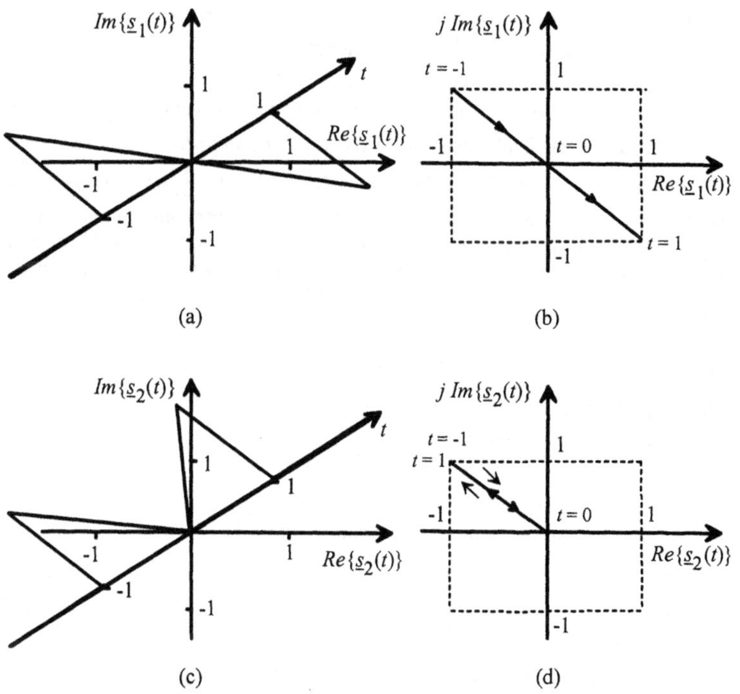

Figure 4-2. Trajectories of complex signals $\underline{s}_1(t)$ and $\underline{s}_2(t)$ for Example 4-1. *3D*-plots of real and imaginary parts as functions of time (*a*), (*c*); *2D*-plots of signal constellations (*b*), (*d*).

It is easy to verify that changing either the upper or the lower limit of integration would make the signals asymmetric relative to the origin and yield a non-zero inner product of the two. If, for instance, we choose the integration interval as [-1, 0], the inner product is

$$\langle \underline{s}_1(t), \underline{s}_2(t) \rangle = \int_{-1}^{0} [t(1-j)][t(1-j)]^* dt = 2\int_{-1}^{0} t^2 dt = -2 \neq 0. \tag{4.27}$$

The signals would then no longer be orthogonal and, of course, would fail to be orthonormal.

▶◀

In the next example, we shall examine how finite-duration complex waveforms $\underline{x}(t)$ can be optimally approximated in a finite-dimensional (say, L-dimensional) vector signal space. There are L orthonormal signals ("unit vectors") available. We shall try to synthesize, as closely as possible, a replica of $\underline{x}(t)$ in the form of a weighted sum (linear combination) of these signals. Wanted are the elements, or coefficients, of the optimal weighting vector $\mathbf{w}_L$. Moreover, should there remain a non-zero error signal $\underline{e}(t)$, we shall calculate its metric and the error energy.

EXAMPLE 4-2: We consider, from the previous example, the complex signals $\underline{\tilde{s}}_1(t)$ and $\underline{\tilde{s}}_2(t)$ of (4.25) which are amplitude-scaled versions of $\underline{s}_1(t)$ and $\underline{s}_2(t)$, respectively, originally given in (4.22). As was proven in Example 4-1, $\underline{\tilde{s}}_1(t)$ and $\underline{\tilde{s}}_2(t)$ constitute a finite-dimensional ($L = 2$) set of orthonormal signals. Now, using linear combinations of these two signals, we wish to approximate as closely as possible, over time interval [-1, 1], an arbitrarily chosen complex waveform. Our finite-duration "test signal" is mathematically described as

$$\underline{x}(t) = \begin{cases} \frac{1}{2}(-t + j) & -1 \le t < 0 \\ \frac{1}{2}(-t + 2j) & 0 \le t \le 1 \\ 0 & else \end{cases} \tag{4.28}$$

Its real and imaginary parts are separately drawn in Figure 4-3.

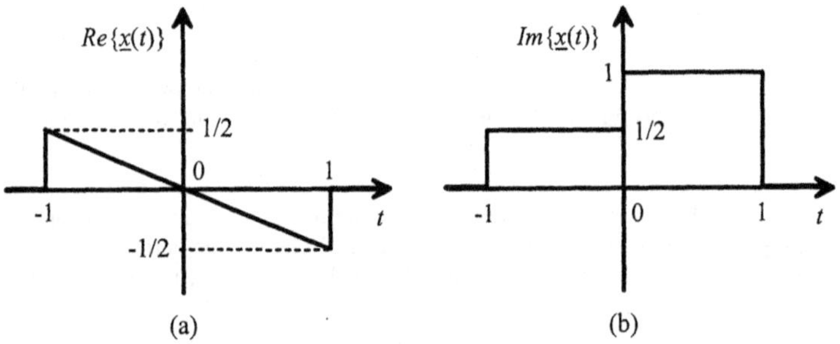

Figure 4-3. Complex test signal $\underline{x}(t)$ for Example 4-2. Plot of real part (a) and imaginary part (b), respectively.

We start by using (4.16) to calculate the two generalized Fourier coefficients as

$$\underline{w}_1 = \int_{-1}^{1} \underline{x}(t)\underline{\tilde{s}}_1^*(t)dt$$
$$= \int_{-1}^{0} [\frac{1}{2}(-t+j)][\frac{1}{2}\sqrt{3}t(1+j)]dt + \int_{0}^{1} [\frac{1}{2}(-t+2j)][\frac{1}{2}\sqrt{3}t(1+j)]dt \tag{4.29}$$
$$= -\frac{1}{24}\sqrt{3}(7+j)$$

and

$$\underline{w}_2 = \int_{-1}^{1} \underline{x}(t)\underline{\tilde{s}}_2^*(t)dt$$
$$= \int_{-1}^{0} [\frac{1}{2}(-t+j)][\frac{1}{2}\sqrt{3}t(1+j)]dt + \int_{0}^{1} [\frac{1}{2}(-t+2j)][-\frac{1}{2}\sqrt{3}t(1+j)]dt \tag{4.30}$$
$$= \frac{3}{8}\sqrt{3}(1-j).$$

Next, by means of (4.17), piecewise approximation (denoted by subscript a) of $\underline{x}(t)$ yields

$$\underline{x}_a(t) = \underline{w}_1\underline{\tilde{s}}_1(t) + \underline{w}_2\underline{\tilde{s}}_2(t) = \begin{cases} -\frac{1}{2}t - j\frac{3}{4}t & -1 \le t < 0 \\ -\frac{1}{2}t + j\frac{3}{2}t & 0 \le t \le 1 \\ 0 & else \end{cases} \tag{4.31}$$

We notice that the error signal

$$\underline{e}(t) = \underline{x}(t) - \underline{x}_a(t) = \begin{cases} j(\frac{1}{2} + \frac{3}{4}t) & -1 \le t < 0 \\ j(1 - \frac{3}{2}t) & 0 \le t \le 1 \\ 0 & else \end{cases} \tag{4.32}$$

is purely imaginary since the real part of $\underline{x}(t)$ is perfectly well (zero error!) approximated by $\underline{x}_a(t)$. Finally, let us calculate the minimized energy contained in the error signal. We found an optimal approximation $\underline{x}_a(t)$ with a minimum error energy of

$$E_x = \langle \underline{e}(t), \underline{e}(t) \rangle = \int_{-1}^{1} \underline{e}(t) \underline{e}^*(t)\, dt$$
$$= \int_{-1}^{0} (\frac{1}{2} + \frac{3}{4} t)^2 dt + \int_{0}^{1} (1 - \frac{3}{2} t)^2 dt = \frac{5}{16} \tag{4.33}$$

and an error metric of

$$\| e(t) \| = \sqrt{E_x} = \frac{1}{4}\sqrt{5}. \tag{4.34}$$

It is left as an additional exercise (see Problem 4.1) to perform, in a straightforward manner, the above steps of piecewise integration, to sketch the real and imaginary parts of $\underline{x}_a(t)$, and to draw a plot of the error signal $\underline{e}(t)$.

►◄

There are a plethora of known classes of orthogonal signal sets. Perhaps the most prominent representatives are complex exponentials observed over a full signal period T.

Suppose we consider, over the time window $[0, T]$, in an infinite-dimensional signal space the set of time-continuous **complex exponentials**

$$\underline{s}_\ell(t) = exp\,(j\,2\pi\,\ell\,t\,/\,T), \quad \ell = 0, \pm 1, \pm 2, \ldots \tag{4.35}$$

We pick out any two of these signals, say $\underline{s}_m(t)$ and $\underline{s}_n(t)$ with arbitrary indices m and n, and proof their orthogonality by calculating their inner product as

$$\langle \underline{s}_m(t), \underline{s}_n(t) \rangle = \int_0^T \underline{s}_m(t)\, \underline{s}_n^*(t)\, dt$$
$$= \int_0^T exp(j\,2\pi\, m\, t\,/\,T) exp(\text{-}j\,2\pi\, n\, t\,/\,T)\, dt \tag{4.36}$$
$$= \begin{cases} T & m = n \\ 0 & m \neq n \end{cases}$$

From there it is easy to conclude that, after normalization, complex exponentials

$$\tilde{\underline{s}}_\ell(t) = \frac{1}{\sqrt{T}} exp(j2\pi\ell t/T), \quad \ell = 0, \pm 1, \pm 2, ... \tag{4.37}$$

represent a set of orthonormal signals over the time interval $[0, T]$.

As an interesting detail, we note that the modulus (or absolute value) of the inner product of two signals is always less or equal to the product of the two signals' norms. Mathematically, we write this in the form of **Schwarz's inequality** as

$$\begin{aligned} \left|\langle \underline{s}_m(t), \underline{s}_n(t)\rangle\right| &\le \|\underline{s}_m(t)\| \, \|\underline{s}_n(t)\| \\ &= \sqrt{\langle \underline{s}_m(t), \underline{s}_m(t)\rangle} \sqrt{\langle \underline{s}_n(t), \underline{s}_n(t)\rangle} \end{aligned} \tag{4.38}$$

Equality exists if and only if signal $\underline{s}_n(t)$ is a weighted version of $\underline{s}_m(t)$, i.e., $\underline{s}_n(t) = \underline{w}\, \underline{s}_m(t)$ where $\underline{w}$ is an arbitrary complex weighting factor.

There are cases where we wish to extend a given set of real-valued orthogonal signals into a set of complex-valued orthogonal signals. Knowing that the real parts are orthogonal to each other, we need to find and add imaginary signal parts such that the resulting complex signals are orthogonal, too. To give an example, let us consider two real signals, $a(t)$ and $c(t)$, with zero inner product over time interval $[t_1, t_2]$. We write the desired orthogonal signals as complex waveforms

$$\underline{s}_1(t) = a(t) + jb(t), \qquad \underline{s}_2(t) = c(t) + jd(t) \tag{4.39}$$

where $b(t)$ and $d(t)$ are the two unknown signals. Then, since we know that

$$\langle a(t), c(t)\rangle = \int_{t_1}^{t_2} a(t)c(t)dt = 0, \tag{4.40}$$

we can make complex signals $\underline{s}_1(t)$ and $\underline{s}_2(t)$ orthogonal to each other by simply choosing their imaginary parts as

$$b(t) = -a(t), \qquad d(t) = -c(t). \tag{4.41}$$

This is readily proofed by calculation of the inner product of the two complex signals. Consideration of (4.40) yields

$$\begin{aligned}\left\langle \underline{s}_1(t), \underline{s}_2(t) \right\rangle &= \int_{t_1}^{t_2} (a(t) - j\,a(t))(c(t) - j\,c(t))^* \,dt \\ &= 2\int_{t_1}^{t_2} a(t)c(t)dt = 0.\end{aligned} \tag{4.42}$$

So far, we dealt with orthogonality of **continuous-time** signals. Practical work with **discrete-time** signals, however, requires a few modifications to be made to the above formulas and mathematical rules. Firstly, we need to replace definite integrals by finite sums and continuous-time signals $\underline{s}(t)$ by their sampled, discrete-time counterparts $\underline{s}(kT_S)$ or, for short, $\underline{s}(k)$ where k is an integer counting index. Sampling time intervals T_S are usually, but not necessarily, being kept constant over time t. Secondly, in digital signal processing, we may retain and even exploit the concept of vector signal spaces. Samples of waveforms are represented by discrete points in a finite-dimensional signal space. Compact notations of multiple discrete-time signals are then possible by writing samples of individual signals in vector form and assembling these vectors as rows or columns, respectively, of signal matrices. Throughout the following text, we shall arrange time samples along the columns of a multiple signal matrices **S**. Suppose we observe N signals in parallel and have available K samples per signal. Then, a **multiple signal matrix** is defined as

$$\mathbf{S} = \begin{pmatrix} \underline{s}_1(1) & \underline{s}_2(2) & \cdots & \underline{s}_N(1) \\ \underline{s}_1(2) & \underline{s}_2(2) & \cdots & \underline{s}_N(2) \\ \vdots & \vdots & \ddots & \vdots \\ \underline{s}_1(K) & \underline{s}_2(K) & \cdots & \underline{s}_N(K) \end{pmatrix} = \left(\mathbf{s}_1, \mathbf{s}_2, \ldots, \mathbf{s}_N\right). \tag{4.43}$$

The nth column (n = 1, 2, ..., N) of matrix **S** includes K sample values taken at discrete times kT_S (k = 1, 2, ..., K) and representing the nth signal. We shall call

$$\mathbf{s}_n = \left(\underline{s}_n(1), \underline{s}_n(2), \ldots, \underline{s}_n(K)\right)^{\mathrm{T}}, \quad n = 1, 2, \ldots, N \tag{4.44}$$

the nth discrete-time signal vector or, for short, **discrete signal vector**.

If we measure, over a finite time interval $[k_1, k_2]$, $\Delta k = k_2 - k_1$ samples of the mth discrete-time complex signal $\underline{s}_m$ and group them consecutively in discrete signal vector $\mathbf{s}_m$, then the energy in $\mathbf{s}_m$ is given by the sum

$$E_m = \sum_{k=k_1}^{k_2} \underline{s}_m(k)\underline{s}_m^*(k) = \sum_{k=k_1}^{k_2} \left|\underline{s}_m(k)\right|^2 . \tag{4.45}$$

In analogy to the properties of continuous-time signals, we may calculate the **inner product** of two discrete signal vectors observed over discrete time interval $[k_1, k_2]$ as a real-valued quantity. It is given by

$$\langle \mathbf{s}_m, \mathbf{s}_n \rangle = \sum_{k=k_1}^{k_2} \underline{s}_m(k)\underline{s}_n^*(k) \qquad 1 \le m,n \le N . \tag{4.46}$$

Any two discrete signal vectors, $\mathbf{s}_m$ and $\mathbf{s}_n$, are **orthogonal** to each other (denoted by $\mathbf{s}_m \perp \mathbf{s}_n$) if their inner product vanishes. In ordinary Euclidean vector space, we would interpret the inner product as the product of the length of vector $\mathbf{s}_m$, the length of vector $\mathbf{s}_n$, and the cosine of the angle between these two vectors. Geometrically, the **natural norm** (symbolically denoted by $\|...\|$) of the mth discrete signal vector $\mathbf{s}_m$ may be interpreted as its length. It is given by the square root of the inner product of $\mathbf{s}_m$ with itself. In other words, we have

$$\begin{aligned} \|\mathbf{s}_m\| &= \sqrt{\langle \mathbf{s}_m, \mathbf{s}_m \rangle} \\ &= \sqrt{E_m} = \sqrt{\sum_{k=k_1}^{k_2} \underline{s}_m(k)\underline{s}_m^*(k)} = \sqrt{\sum_{k=k_1}^{k_2} \left|\underline{s}_m(k)\right|^2} . \end{aligned} \tag{4.47}$$

Following the above interpretation of signals as vectors in Euclidean space and using direction cosines, the angle $\varphi_{m,n}$ between discrete signal vectors $\mathbf{s}_m$ and $\mathbf{s}_n$ is obtained as the inverse cosine given by

$$\begin{aligned} \phi_{m,n} &= cos^{-1}\left(\frac{\langle \mathbf{s}_m, \mathbf{s}_n \rangle}{\|\mathbf{s}_m\|\|\mathbf{s}_n\|} \right) \\ &= cos^{-1}\left(\frac{\sum_{k=k_1}^{k_2} \underline{s}_m(k)\underline{s}_n^*(k)}{\sqrt{\left(\sum_{k=k_1}^{k_2} \left|\underline{s}_m(k)\right|^2\right)\left(\sum_{k=k_1}^{k_2} \left|\underline{s}_n(k)\right|^2\right)}} \right) \end{aligned} \tag{4.48}$$

Suppose $\mathbf{s}_m$ and $\mathbf{s}_n$ are both vectors composed of K real samples (K-tuples). Then we can numerically determine the inner product by calculating the dot product known from vector algebra as

$$\langle \mathbf{s}_m, \mathbf{s}_n \rangle = \mathbf{s}_m^T \cdot \mathbf{s}_n . \tag{4.49}$$

For complex signal vectors, one replaces $\mathbf{s}_m^T$ by the Hermetian of complex vector $\mathbf{s}_m$ which is denoted by $\mathbf{s}_m^H$.

Furthermore, for a set of discrete signal vectors to be orthonormal, we require that all pairs of discrete signal vectors in that set have unit energy. Two orthogonal discrete-time signals, s_m and $\mathbf{s}_n$, can be made orthonormal over sample time interval $[k_1, k_2]$ by proper scaling, as we did with continuous-time signals. Suppose energies E_m and/or E_n are unequal one. The inner product of the corresponding **orthonormalized** signals (superscripts tilde = both vectors have unit norm) must then be of the form

$$\langle \tilde{\mathbf{s}}_m, \tilde{\mathbf{s}}_n \rangle = \sum_{k=k_1}^{k_2} \left(\frac{s_m(k)}{\sqrt{E_m}} \cdot \frac{s_n^*(k)}{\sqrt{E_n}} \right) = \begin{cases} 1, & m = n \\ 0, & m \neq n \end{cases} \tag{4.50}$$

$$= \delta(m - n)$$

where symbol δ stands for the Kronecker delta function.

We shall now stop for a while and look at an example of how to apply the above definitions to a low-dimensional vector space. To simplify the numerical problems as much as possible, we shall choose an R^2 vector space spanned by a set of two nonorthogonal real (2×1) column vectors.

EXAMPLE 4-3: Suppose we have available very short snapshots of two ($N = 2$) real signals, each one represented by no more than two ($K = 2$) samples. Using our multiple signal matrix notation, these signals are given as

$$\mathbf{S} = \begin{pmatrix} 1 & 3 \\ 2 & 1 \end{pmatrix} = \begin{pmatrix} s_{1,1} & s_{2,1} \\ s_{1,2} & s_{2,2} \end{pmatrix} = (\mathbf{s}_1, \mathbf{s}_2) . \tag{4.51}$$

In a two-dimensional cartesian coordinate system ($z = 0$) with orthogonal unit vectors $\mathbf{u}_x = \mathbf{u}_1 = (1, 0)^{\mathrm{T}}$ and $\mathbf{u}_y = \mathbf{u}_2 = (0, 1)^{\mathrm{T}}$, respectively, we can draw signal vectors $\mathbf{s}_1$ and $\mathbf{s}_2$ as shown in Figure 4-4. Both of them are directed lines ranging from the origin $(0, 0)^{\mathrm{T}}$ to the points with coordinates specified as multiples of these unit vectors. It is readily seen that the lengths of the given signal vectors can be viewed as the square roots of inner products

$$\|\mathbf{s}_1\| = \sqrt{\langle \mathbf{s}_1, \mathbf{s}_1 \rangle} = \sqrt{s_{1,1}^2 + s_{1,2}^2} = \sqrt{5}$$
$$\|\mathbf{s}_2\| = \sqrt{\langle \mathbf{s}_2, \mathbf{s}_2 \rangle} = \sqrt{s_{2,1}^2 + s_{2,2}^2} = \sqrt{10} \tag{4.52}$$

The angle between discrete signal vectors $\mathbf{s}_1$ and $\mathbf{s}_2$ is given by

$$\phi = \phi_{1,2} = \angle(\mathbf{s}_1 - \mathbf{s}_2) = cos^{-1}(\frac{\langle \mathbf{s}_1, \mathbf{s}_2 \rangle}{\|\mathbf{s}_1\| \|\mathbf{s}_2\|}) = cos^{-1}(\frac{5}{5\sqrt{2}}) = 45° \tag{4.53}$$

and the distance from the tip of $\mathbf{s}_1$ to the tip of vector $\mathbf{s}_2$ is equal to the norm of difference vector ($\mathbf{s}_2$ - $\mathbf{s}_1$) or, in terms of the given sample amplitudes,

$$\|\mathbf{s}_2 - \mathbf{s}_1\| = \sqrt{(s_{2,1} - s_{1,1})^2 + (s_{2,2} - s_{1,2})^2} = \sqrt{5}\,. \tag{4.54}$$

Another aspect well known from plane geometry is that we can express $\mathbf{s}_1$ as the vector sum of a vector $\mathbf{s}_{1|\mathbf{s}_2}$, i.e., collinear with $\mathbf{s}_2$ and a vector orthogonal to $\mathbf{s}_2$, termed $\mathbf{s}_{1\perp\mathbf{s}_2}$ in the following. The lengths of vectors $\mathbf{s}_{1|\mathbf{s}_2}$ and $\mathbf{s}_{1\perp\mathbf{s}_2}$ are represented by letters a and b, respectively, in Figure 4-4. Similar arguments hold if $\mathbf{s}_2$ is projected onto signal vector $\mathbf{s}_1$.

Thus, we may write $\mathbf{s}_1$ and $\mathbf{s}_2$ as

$$\mathbf{s}_1 = \mathbf{s}_{1|\mathbf{s}_2} + \mathbf{s}_{1\perp\mathbf{s}_2} \; (= \mathbf{a} + \mathbf{b})$$
$$\mathbf{s}_2 = \mathbf{s}_{2|\mathbf{s}_1} + \mathbf{s}_{2\perp\mathbf{s}_1} \tag{4.55}$$

where the projection vectors are calculated as

$$\mathbf{s}_{1|\mathbf{s}_2} = \frac{\langle \mathbf{s}_1, \mathbf{s}_2 \rangle}{\|\mathbf{s}_2\|^2} \mathbf{s}_2 = \frac{1}{2}(3,1)^{\mathrm{T}} \quad (= \mathbf{a})$$
$$\mathbf{s}_{2|\mathbf{s}_1} = \frac{\langle \mathbf{s}_1, \mathbf{s}_2 \rangle}{\|\mathbf{s}_1\|^2} \mathbf{s}_1 = (1, 2)^{\mathrm{T}} \tag{4.56}$$

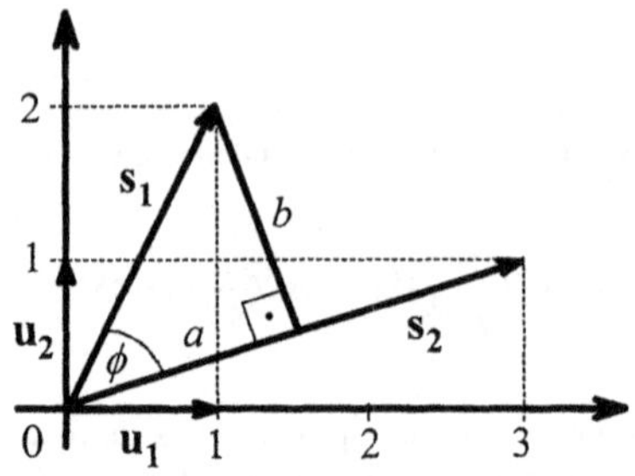

Figure 4-4. Discrete signal vectors $\mathbf{s}_1$ and $\mathbf{s}_2$ for Example 4-3. Projection of $\mathbf{s}_1$ onto $\mathbf{s}_2$ (i.e., $\mathbf{s}_{1|\mathbf{s}_2}$, collinear with $\mathbf{s}_2$) shown as $a = \|\mathbf{a}\|$, and vector $\mathbf{s}_{1\perp\mathbf{s}_2} = \mathbf{b} = \mathbf{s}_1 - \mathbf{a} = \mathbf{s}_1 - \mathbf{s}_{1|\mathbf{s}_2}$ represented by its length $b = \|b\|$. *Note*: Coincidentally, the projection of $\mathbf{s}_2$ onto $\mathbf{s}_1$ is equal to $\mathbf{s}_1$ itself.

Next, we note that vectors $\mathbf{s}_{1\perp\mathbf{s}_2}$ and $\mathbf{s}_{2\perp\mathbf{s}_1}$ are orthogonal to the projection vectors $\mathbf{s}_{1|\mathbf{s}_2}$ and $\mathbf{s}_{1\perp\mathbf{s}_2}$. Identification of the lengths and directions of $\mathbf{s}_{1\perp\mathbf{s}_2}$ and $\mathbf{s}_{2\perp\mathbf{s}_1}$ is a two-step task. First, their orientations must be chosen perpendicular to $\mathbf{s}_2$ and $\mathbf{s}_1$, respectively. In a second step, their norms (lengths) need to be properly adjusted. Rotations of vectors $\mathbf{s}_1$ by +90° and and $\mathbf{s}_2$ by -90°, respectively, yield the vectors orthogonal to $\mathbf{s}_1$ and $\mathbf{s}_2$ (prior to length correction!) as

$$\begin{aligned} \hat{\mathbf{s}}_{1\perp\mathbf{s}_2} &= (-s_{2,2}, s_{2,1})^T = (-1, 3)^T \\ \hat{\mathbf{s}}_{2\perp\mathbf{s}_1} &= (s_{1,2}, -s_{1,1})^T = (2, -1)^T \end{aligned} \tag{4.57}$$

Length scaling is then performed by putting a scalar factor in front of these two vectors. Using plane geometry, we find the correctly scaled vector quantities as

$$\begin{aligned} \mathbf{s}_{1\perp\mathbf{s}_2} &= \frac{\langle \mathbf{s}_1, \mathbf{s}_2 \rangle}{\|\mathbf{s}_2\|^2} \hat{\mathbf{s}}_{1\perp\mathbf{s}_2} = \frac{1}{2}(-1, 3)^T \quad (= \mathbf{b}) \\ \mathbf{s}_{2\perp\mathbf{s}_1} &= \frac{\langle \mathbf{s}_1, \mathbf{s}_2 \rangle}{\|\mathbf{s}_1\|^2} \hat{\mathbf{s}}_{2\perp\mathbf{s}_1} = (2, -1)^T \end{aligned} \tag{4.58}$$

The reader should check the results by means of (4.55) with the two closed loops (right triangles) involved, each one forming the sum of three vectors.

It is trivial here, yet instructive, to express matrix **S** as the product of unit matrix **U** (columns consist of orthonormal standard basis vectors $\mathbf{u}_1$ and $\mathbf{u}_2$) and an appropriately chosen weight matrix **W**. We may write our given multiple signal matrix in the form of

$$\mathbf{S} = \mathbf{W} \cdot \mathbf{U} = \begin{pmatrix} w_{1,1} & w_{2,1} \\ w_{1,2} & w_{2,2} \end{pmatrix} \cdot \begin{pmatrix} 1 & 0 \\ 0 & 1 \end{pmatrix} = (\mathbf{w}_1, \mathbf{w}_2) \cdot (\mathbf{u}_1, \mathbf{u}_2) . \tag{4.59}$$

Of course, $\mathbf{S} = \mathbf{W}$ in this example but, as we have already seen in the context of continuous-time signals, any other choice of the set of orthonormal basis vectors would require nontrivial calculations of a different weighting matrix, i.e., $\mathbf{W} \neq \mathbf{S}$. For instance, choosing the two orthonormal basis vectors as

$$\left.\begin{matrix} \mathbf{u}_1 = \frac{1}{2}\sqrt{2}(1, -1)^{\mathrm{T}} \\ \mathbf{u}_1 = \frac{1}{2}\sqrt{2}(1, \; 1)^{\mathrm{T}} \end{matrix}\right\} \Leftrightarrow \mathbf{U} = \frac{1}{2}\sqrt{2}\begin{pmatrix} 1 & 1 \\ -1 & 1 \end{pmatrix} \tag{4.60}$$

would yield a weighting matrix of

$$\mathbf{W} = \mathbf{S} \cdot \mathbf{U}^{-1} = \frac{1}{2}\sqrt{2}\begin{pmatrix} 1 & 3 \\ 2 & 1 \end{pmatrix}\begin{pmatrix} 1 & -1 \\ 1 & 1 \end{pmatrix} = \frac{1}{2}\sqrt{2}\begin{pmatrix} 4 & 2 \\ 3 & -1 \end{pmatrix}. \tag{4.61}$$

►◄

4.4 SIGNAL PROJECTIONS

The results obtained in Example 4-3 can be generalized to higher-dimensional inner product spaces. In fact, any finite-dimensional inner product space has, as its possible bases, sets of orthonormal vectors. To find, for a given $(K \times N)$ signal matrix **S**, a set of orthonormal basis vectors, we note that any discrete signal vector $\mathbf{s}_m$ can be projected onto one out of a set of **orthogonal** basis vectors. If, for instance, we wish to project $\mathbf{s}_m$ onto vector $\tilde{\mathbf{u}}_n$ (where, to distinguish ortho*gonal* from ortho*normal* vectors, superscript tilde (~) stands for a nonnormalized vector), the projection vector is collinear with $\tilde{\mathbf{u}}_n$. Using the **projection theorem**, the projection of $\mathbf{s}_m$ onto $\tilde{\mathbf{u}}_n$ can be calculated as

$$proj(\mathbf{s}_m, \tilde{\mathbf{u}}_n) = \mathbf{s}_{m|\tilde{\mathbf{u}}_n} = \frac{\langle \mathbf{s}_m, \tilde{\mathbf{u}}_n \rangle}{\|\tilde{\mathbf{u}}_n\|^2}\tilde{\mathbf{u}}_n . \tag{4.62}$$

Suppose our discrete signal matrix **S** consists of N linearly independent nonorthogonal column vectors, $\mathbf{s}_1$, $\mathbf{s}_2$, ... $\mathbf{s}_N$, each one with K samples or observations, s_1, s_2, ..., s_K. Now, by using the projection theorem, we can produce an orthonormal basis in a recursive manner. Starting with vector $\mathbf{s}_1$, we first calculate from $\mathbf{S}$ a matrix $\widetilde{\mathbf{U}}$ whose columns are mutually ortho*gonal* basis vectors. An ortho*normal* basis $\mathbf{U}$ (one that spans the column space of $\mathbf{S}$) is then obtained from $\widetilde{\mathbf{U}}$ by normalizing the column vectors of $\widetilde{\mathbf{U}}$. The algorithm is known as the "classical" **Gram-Schmidt procedure** (named after mathematicians J. P. Gram and Erhard Schmidt). Basically, the orthogonalization steps are as follows:

$$\widetilde{\mathbf{u}}_1 = \mathbf{s}_1 , \tag{4.63}$$

$$\begin{aligned}\widetilde{\mathbf{u}}_n &= \mathbf{s}_n - \sum_{i=1}^{n-1} proj(\mathbf{s}_n, \widetilde{\mathbf{u}}_i) \\ &= \mathbf{s}_n - \sum_{i=1}^{n-1} \frac{\langle \mathbf{s}_n, \widetilde{\mathbf{u}}_i \rangle}{\|\widetilde{\mathbf{u}}_i\|^2} \widetilde{\mathbf{u}}_i, \quad n = 2, 3, ..., N.\end{aligned} \tag{4.64}$$

Following these recusions, normalization of each one of the columns of $\widetilde{\mathbf{U}}$ yields the column vectors of $\mathbf{U}$, the orthonormal basis, as

$$\mathbf{u}_n = \frac{1}{\|\widetilde{\mathbf{u}}_n\|} \widetilde{\mathbf{u}}_n, \quad n = 1, 2, ..., N. \tag{4.65}$$

It should be mentioned that various extensions and numerically more stable modifications of the classical Gram-Schmidt orthonormalization procedure exist [1]. Note also that

$$\mathbf{U} \cdot \mathbf{U}^T = \mathbf{I} . \tag{4.66}$$

Recursion (4.64) can be written as

$$\begin{aligned}\mathbf{s}_n &= \widetilde{\mathbf{u}}_n + \sum_{i=1}^{n-1} \frac{\langle \mathbf{s}_n, \widetilde{\mathbf{u}}_i \rangle}{\|\widetilde{\mathbf{u}}_i\|^2} \widetilde{\mathbf{u}}_i \\ &= \widetilde{\mathbf{u}}_n + \sum_{i=1}^{n-1} r_{n,i} \widetilde{\mathbf{u}}_i, \quad n = 2, 3, ..., N.\end{aligned} \tag{4.67}$$

In a columnwise notation, we may hence express our $(N \times K)$ signal matrix **S** as

$$\mathbf{S} = (\mathbf{s}_1, \mathbf{s}_2, \cdots, \mathbf{s}_N)$$
$$= (\tilde{\mathbf{u}}_1, \tilde{\mathbf{u}}_2, \cdots, \tilde{\mathbf{u}}_N) \cdot \begin{pmatrix} 1 & r_{2,1} & r_{3,1} & \cdots & r_{N,1} \\ 0 & 1 & r_{3,2} & \cdots & r_{N,2} \\ \vdots & \vdots & \vdots & \ddots & \vdots \\ 0 & 0 & 0 & \cdots & 1 \end{pmatrix} \tag{4.68}$$

It is customary in linear algebra to symbolically denote $(K \times N)$ orthogonal matrix $\tilde{\mathbf{U}}$ by capital letter **Q**, and to use capital letter **R** for the **square** $(N \times N)$, **upper triangular**, **invertible** matrix on the right side of equation (4.68). To summarize our conclusions from the Gram-Schmidt orthonormalization procedure, we may say that any $(K \times N)$ signal matrix **S** with linearly independent columns can be factorized into a matrix product of the form

$$\mathbf{S} = \mathbf{Q} \cdot \mathbf{R} . \tag{4.69}$$

With the extra requirement of making the first nonzero element in each row of **R** nonnegative (which is always possible!), the above decomposition of **S** is **unique**. Equation (4.69) is known as *QR*-decomposition. It plays an important role not only in linear algebra but also in various areas of applied signal processing such as least squares estimation, channel identification, equalization and optimization of filter weights, etc. The basic MATLAB software package offers, for a given $(K \times N)$ signal matrix **S**, a numerically robust *QR*-decomposition algorithm in the form of a single command line

[Q, R] = qr(S) (4.70)

Note that MATLAB, besides making **Q** an orthogonal basis, performs an additional normalization step on each one of the column vectors. Therefore, what you get is an orthonormal basis **Q**, i.e., one that is identical to our definition of an orthonormal basis **U**. To compensate for that unit vector length scaling, the elements of matrix **R** are reciprocally scaled in the course of MATLAB's *QR*-decomposition procedure.

We proceed now with a numerical example before showing a few technical applications of orthonormalization algorithms.

EXAMPLE 4-4: We use again our simple (2×2) signal matrix **S** from the previous excercise. Recall from Example 4-3 that the columns of **S** are given by $\mathbf{s}_1 = (1, 2)^T$ and $\mathbf{s}_2 = (3, 1)^T$, respectively. The first task here is to apply the classical Gram-Schmidt orthogonalization procedure to obtain a (2×2) matrix **U** with mutually orthogonal columns of unit norm. Following the recursion steps of (4.63) and (4.64), we have

$$\tilde{\mathbf{u}}_1 = \mathbf{s}_1 = \begin{pmatrix} 1 \\ 2 \end{pmatrix}, \quad \|\tilde{\mathbf{u}}_1\|^2 = \langle \tilde{\mathbf{u}}_1, \tilde{\mathbf{u}}_1 \rangle^2 = \tilde{\mathbf{u}}_1^T \cdot \tilde{\mathbf{u}}_1 = 5, \tag{4.71}$$

$$proj(\mathbf{s}_2, \tilde{\mathbf{u}}_1) = \mathbf{s}_{2|\tilde{\mathbf{u}}_1} = \frac{\langle \mathbf{s}_2, \tilde{\mathbf{u}}_1 \rangle}{\|\tilde{\mathbf{u}}_1\|^2} \tilde{\mathbf{u}}_1 = \frac{5}{5}\begin{pmatrix} 1 \\ 2 \end{pmatrix} = \begin{pmatrix} 1 \\ 2 \end{pmatrix}, \tag{4.72}$$

$$\tilde{\mathbf{u}}_2 = \mathbf{s}_2 - proj(\mathbf{s}_2, \tilde{\mathbf{u}}_1) = \begin{pmatrix} 2 \\ -1 \end{pmatrix}, \quad \|\tilde{\mathbf{u}}_2\| = \sqrt{\langle \tilde{\mathbf{u}}_2, \tilde{\mathbf{u}}_2 \rangle} = \sqrt{\tilde{\mathbf{u}}_2^T \cdot \tilde{\mathbf{u}}_2} = \sqrt{5}\,. \tag{4.73}$$

Note that column vectors $\tilde{\mathbf{u}}_1$ and $\tilde{\mathbf{u}}_2$ of matrix $\tilde{\mathbf{U}}$ are mutually orthogonal but not orthonormal, hence the superscript ~. Divisions by their respective norms are required to obtain the orthonormal basis **U** as

$$\mathbf{U} = (\mathbf{u}_1, \mathbf{u}_2) = (\frac{1}{\|\tilde{\mathbf{u}}_1\|}\tilde{\mathbf{u}}_1, \frac{1}{\|\tilde{\mathbf{u}}_2\|}\tilde{\mathbf{u}}_2) = \frac{1}{5}\sqrt{5}\begin{pmatrix} 1 & 2 \\ 2 & -1 \end{pmatrix}. \tag{4.74}$$

It is readily shown that column vectors $\mathbf{u}_1$ and $\mathbf{u}_2$ are of unit length and that their inner product vanishes, i.e.,

$$\langle \mathbf{u}_1, \mathbf{u}_2 \rangle = \mathbf{u}_1^T \cdot \mathbf{u}_2 = 0\,. \tag{4.75}$$

We may express **S** in the orthonormal vector space spanned by **U** as a weighted sum of $\mathbf{u}_1$ and $\mathbf{u}_2$, respectively. Hence, for a linear combination $\mathbf{S} = \mathbf{W} \cdot \mathbf{U}$, we calculate weighting matrix **W** as

$$\mathbf{W} = \mathbf{S} \cdot \mathbf{U}^{-1} = \frac{1}{5}\sqrt{5}\begin{pmatrix} 7 & -1 \\ 4 & 3 \end{pmatrix}. \tag{4.76}$$

In the next step, we shall see how the *QR*-decomposition algorithm can be applied to factorize the given signal matrix **S** such that

$$\mathbf{S} = \mathbf{Q} \cdot \mathbf{R}\,. \tag{4.77}$$

where the columns of (2×2) matrix $\mathbf{Q}$ are mutually orthogonal and of unit norm, and $\mathbf{R}$ is an upper triangular, invertible matrix. We write $\mathbf{S}$ columnwise in the form of

$$\mathbf{S} = (\mathbf{s}_1, \mathbf{s}_2) = (\tilde{\mathbf{u}}_1, \tilde{\mathbf{u}}_2)\cdot\begin{pmatrix} 1 & r_{2,1} \\ 0 & 1 \end{pmatrix} \tag{4.78}$$

where, according to equation (4.67) and with an upper summation index of n -1 = 1, element $r_{2,1}$ of matrix $\mathbf{R}$ can be obtained from

$$\mathbf{s}_2 = \tilde{\mathbf{u}}_2 + \frac{\langle \mathbf{s}_2, \tilde{\mathbf{u}}_1\rangle}{\|\tilde{\mathbf{u}}_1\|^2}\tilde{\mathbf{u}}_1 = \tilde{\mathbf{u}}_2 + r_{2,1}\tilde{\mathbf{u}}_1 = \begin{pmatrix} 2 \\ -1 \end{pmatrix} + r_{2,1}\begin{pmatrix} 1 \\ 2 \end{pmatrix} = \begin{pmatrix} 3 \\ 1 \end{pmatrix}. \tag{4.79}$$

Thus, we find $r_{2,1} = 1$, and $\mathbf{S}$ can be QR-factorized as

$$\mathbf{S} = \mathbf{Q}\cdot\mathbf{R} = \begin{pmatrix} 1 & 2 \\ 2 & -1 \end{pmatrix}\cdot\begin{pmatrix} 1 & 1 \\ 0 & 1 \end{pmatrix} = \frac{1}{\sqrt{5}}\begin{pmatrix} 1 & 2 \\ 2 & -1 \end{pmatrix}\cdot\sqrt{5}\begin{pmatrix} 1 & 1 \\ 0 & 1 \end{pmatrix}. \tag{4.80}$$

Note in particular the right-most part of (4.80) as MATLAB will output matrices $\mathbf{Q}$ and $\mathbf{R}$ in that format. The reader is encouraged to repeat the various steps above and to verify the results of QR-decomposition by means of no more than two lines of MATLAB code.

▶◀

4.5 CHAPTER 4 PROBLEMS

4.1 For the complex test signal $\underline{x}(t)$ of Example 4-2, perform the intermediate steps of calculating the two-dimensional weight vector $\mathbf{w}_2 = (\underline{w}_1, \underline{w}_2)^T$, the approximated signal $\underline{x}_a(t)$, error signal $\underline{e}(t)$, and the error energy E_x. Make plots of the real and imaginary parts of all signals involved.

4.2 Apply the Gram-Schmidt orthonormalization procedure to a square signal matrix with the following real elements: $\mathbf{S} = \begin{pmatrix} 1 & 2 & 4 \\ 4 & 3 & 2 \\ -1 & 1 & 2 \end{pmatrix}$. How can $\mathbf{S}$ be QR-factorized into a matrix product of the form $\mathbf{S} = \mathbf{Q}\cdot\mathbf{R}$ where both $\mathbf{Q}$ and $\mathbf{R}$ are (3×3) matrices. Use MATLAB to verify your results.

REFERENCE

[1] G. H. Golub and C. F. Van Loan. *Matrix Computations*. The Johns Hopkins University Press, Baltimore, Md, 3rd edition, 1996.

Chapter 5

MIMO CHANNEL PROBLEMS AND SOLUTIONS

5.1 INTRODUCTION

QR-decomposition can be helpful in various technical applications where computational complexity is an issue. Keeping the number of multiplications and additions down to a minimum is one of the biggest challenges in real-time signal processing. Let us now consider some typical problems where factorization of matrices and analyses in vector signal space can significantly reduce the number of arithmetic calculations.

5.2 STANDARD PROBLEMS ENCOUNTERED IN MIMO SYSTEM THEORY

First, we study a **noise-free** MIMO-type transmission channel shown in Figure 5-1 (*a*) with N inputs and M outputs. Real or complex-valued signals are measured at each one of these ports. We observe all of them in parallel at discrete multiples ($k = 1, 2, 3, ...$) of equidistant sample time intervals of T_s seconds. Let us group blocks of K observations per input into a K-by-N input signal matrix $\mathbf{X} = (\mathbf{x}_1, \mathbf{x}_2, ..., \mathbf{x}_N)$. Each column of $\mathbf{X}$ contains N samples and belongs to a distinct input port. Similarly, we observe K discrete samples at the M outputs and define $\mathbf{Y} = (\mathbf{y}_1, \mathbf{y}_2, ... , \mathbf{y}_M)$ as the output signal matrix. Updates of $\mathbf{X}$ and $\mathbf{Y}$ become available every KT_s seconds. We shall count these update intervals (in terms of multiples of K) by using integer numbers $p = K, 2K, 3K, ...$

Let us further assume the channel exhibits **time-varying** behavior, i.e., its transfer characteristics may change dynamically from one set of observations to the next, i.e.,.from p to $p+1$, and so on. Consequently, the outputs will be linked to the inputs by a real or complex matrix of weighting factors or gains, $\mathbf{W}_c$. All elements of matrix $\mathbf{W}_c$ may change over time.

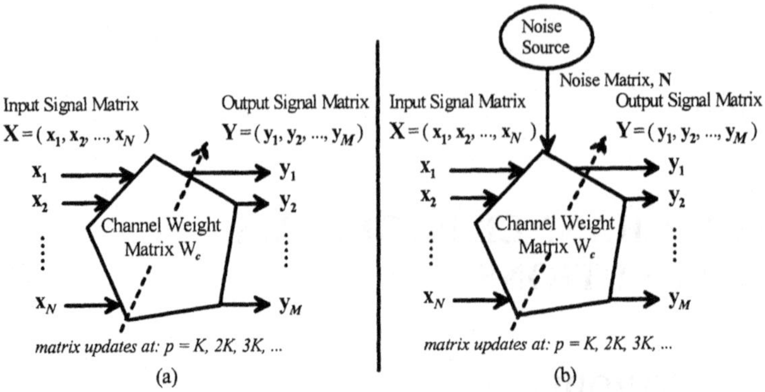

Figure 5-1. System models of time-varying MIMO channels with N inputs and M outputs observed at discrete sample times $k = 1, 2, ..., K$. Updates of channel weight matrix $\mathbf{W}_c$ are computed at discrete times $p = K, 2K, 3K, ...$ (*a*) Noise-free channel; (*b*) channel with noise.

The channel hence represents a time-varying system. As we observe N inputs and M outputs, each one over K uniformly spaced discrete instants of time, the appropriate size of the channel weight matrix $\mathbf{W}_c$ should be $(N \times M)$. In a non-overlapping mode and based on full blocks of completely new input/output samples, new estimates of $\mathbf{W}_c$ can only be computed after every K sample time intervals T_s. The **maximum change rate** of the channel characteristics should, therefore, not exceed

$$f_{\mathbf{W}_c}^{(max)} = \frac{1}{2KT_s}. \tag{5.1}$$

For the **deterministic** noise-free case (see Figure 5-1 (*a*)) we may ask three standard questions:

(1a) Given $\mathbf{Y}$ and $\mathbf{W}_c$, what has been sent over the channel? Determine $\mathbf{X}$.

(2a) Knowing $\mathbf{X}$ and $\mathbf{Y}$, how did the channel transform inputs $\mathbf{X}$ into outputs $\mathbf{Y}$ at discrete sample times $k = 1, 2, ...K$? Identify $\mathbf{W}_c$.

(3a) Given $\mathbf{X}$ and $\mathbf{W}_c$, what is the channel output $\mathbf{Y}$? Calculate $\mathbf{Y}$.

To answer any of these questions, the **noise-free** channel behavior is first modeled by a $(K \times N) \cdot (N \times M)$ matrix product in the form of

$$\mathbf{Y} = \mathbf{X} \cdot \mathbf{W}_c \tag{5.2}$$

and then solved for the missing matrix. Thus, we distinguish between solutions to

MIMO Channel Problem 1a: $\hat{\mathbf{X}} = f_{1a}(\mathbf{W}_c, \mathbf{Y}), \mathbf{N} = \mathbf{0},$ (5.3)

MIMO Channel Problem 2a: $\hat{\mathbf{W}}_c = f_{2a}(\mathbf{X}, \mathbf{Y}), \mathbf{N} = \mathbf{0},$ (5.4)

MIMO Channel Problem 3a: $\hat{\mathbf{Y}} = f_{3a}(\mathbf{W}_c, \mathbf{X}), \mathbf{N} = \mathbf{0}$ (5.5)

where f, with the appropriate subscript, stands for a matrix function and superscript hat (^) symbolically denotes an estimate of the respective matrix.

In a similar manner, we shall categorize the **stochastic** cases (see Figure 5-1 (*b*)) where noise (in any statistical distribution) is present on the channel's paths and hence, to an unknown extent, corrupts the output signals. As we don't know the elements of noise matrix **N**, we are left with three more complicated questions. To complete our taxonomie of technical problems in the context of time-varying MIMO channels, we shall refer to these estimation problems as

MIMO Channel Problem 1b:

$$\hat{\mathbf{X}} = f_{1b}(\mathbf{W}_c, \mathbf{Y}),\ \mathbf{N} \neq \mathbf{0}, \quad \text{subject to } \min_{\mathbf{W}_c, \mathbf{Y}} \left\| \mathbf{X} - \hat{\mathbf{X}} \right\|_F, \tag{5.6}$$

MIMO Channel Problem 2b:

$$\hat{\mathbf{W}}_c = f_{2b}(\mathbf{X}, \mathbf{Y}),\ \mathbf{N} \neq \mathbf{0}, \quad \text{subject to } \min_{\mathbf{X}, \mathbf{Y}} \left\| \mathbf{W}_c - \hat{\mathbf{W}}_c \right\|_F, \tag{5.7}$$

MIMO Channel Problem 3b:

$$\hat{\mathbf{Y}} = f_{3b}(\mathbf{W}_c, \mathbf{X}),\ \mathbf{N} \neq \mathbf{0}, \quad \text{subject to } \min_{\mathbf{W}_c, \mathbf{X}} \left\| \mathbf{Y} - \hat{\mathbf{Y}} \right\|_F \tag{5.8}$$

In the above problem descriptions, we use $\left\| \mathbf{S} \right\|_F$ as the **Frobenius norm** of some real or complex signal matrix **S** with K rows and N columns. It is defined by

$$\|\mathbf{S}\|_F = \sqrt{\sum_{k=1}^{K}\sum_{n=1}^{N}\left|\underline{s}_{k,n}\right|^2} \tag{5.9}$$

where $\underline{s}_{k,n}$ is the k,nth element of **S**. Of course, this definition of the Frobenius norm also applies to a real or complex channel matrix $\mathbf{W}_c$ whose dimensions are N by M.

Problems of type 1b are of fundamental importance in receiver design where **X** represents the transmitted data symbols (often symbolically denoted by the matrix of code words, **C**) and estimates thereof are to be detected in the form of signal matrix $\hat{\mathbf{X}}$. The main goal in **symbol detection** is to minimize, under noisy channel conditions, the average error energy expressed in terms of the squared norm of matrix difference $\mathbf{X}-\hat{\mathbf{X}}$.

Equally important and a prerequisite in the process of channel equalization is the problem of **channel identification** which can be found on the above list under category 2b. The task there is to obtain a mathematical channel model $\hat{\mathbf{W}}_c$ that most closely resembles the true channel behavior as expressed by the desired (correct) matrix of channel weights $\mathbf{W}_c$.

Finally, Problem 3b encompasses the errors in the **measurement** of output signal matrix **Y** induced by channel noise. Even if we had perfect information about both the transmitted symbols **X** and the noise-free channel matrix $\mathbf{W}_c$, the received signals $\hat{\mathbf{Y}}$ are corrupted by noise, $\mathbf{N} \neq \mathbf{0}$. Therefore, our observations $\hat{\mathbf{Y}}$ are nothing else but snapshots taken of a noise process. If we are lucky, we know something about the statistical parameters (mean, standard deviation, and/or higher order moments) of that process. Anyway, we are inevitably faced with situations where $\hat{\mathbf{Y}} \neq \mathbf{X}\cdot\mathbf{W}_c$. In a K-dimensional signal space, the tips of the received (column) vectors $\hat{\mathbf{y}}_m$, m = 1, 2, ... M, tend to deviate in an unpredictable fashion from their ideal positions $\mathbf{y}_m$, thereby producing a noise-induced "cloud" of discrete points in that vector space.

To find out more about the practical advantages of vector space representations of signals, let us investigate firstly the noise-free cases listed under *MIMO Channel Problems 1a* through *3a*. As *Problem 3a* is readily solved by a simple matrix multiplication ($\mathbf{Y}=\mathbf{X}\cdot\mathbf{W}_c$, $\mathbf{N}=\mathbf{0}$), we shall focus below on Problems 1a and 2a, respectively.

Provided that the MIMO system is not ill-conditioned (i.e., $det(\mathbf{W}_c) \neq 0$) and if $\mathbf{W}_c$ is a square matrix (i.e., number of inputs equals number of system outputs, N = M), *MIMO Channel Problem 1a* can then be solved in a elementary manner known from linear algebra by calculation of the inverse of the square ($N \times N$) weight matrix $\mathbf{W}_c$ and right-multiplication of (5.2) by $\mathbf{W}_c^{-1}$. The exact solution of input signal matrix **X** is then obtained as

$$\hat{\mathbf{X}} = \mathbf{X} = \mathbf{Y} \cdot \mathbf{W}_c^{-1}. \tag{5.10}$$

In practical applications, however, the number of channel outputs may be different from the number of inputs. The channel matrix $\mathbf{W}_c$ under consideration then takes on a rectangular ($N \times M$) format, and an appropriate inverse of $\mathbf{W}_c$ needs to be calculated. For <u>real</u> matrices, we replace $\mathbf{W}_c^{-1}$ by the **Moore-Penrose pseudoinverse** given by

$$\mathbf{W}_c^{+} = \begin{cases} (\mathbf{W}_c^T \cdot \mathbf{W}_c)^{-1} \cdot \mathbf{W}_c^T, & M < N \\ \mathbf{W}_c^T \cdot (\mathbf{W}_c \cdot \mathbf{W}_c^T)^{-1}, & M > N \end{cases} \tag{5.11}$$

Using $\mathbf{W}_c^{+}$ instead of $\mathbf{W}_c^{-1}$, as in (5.10), yields the best one of all possible solutions $\hat{\mathbf{X}}$ of MIMO Channel Problem 1a (5.3). The optimal solution $\hat{\mathbf{X}}_{opt}$ is calculated from **Y** and the pseudoinverse of $\mathbf{W}_c$ as

$$\hat{\mathbf{X}}_{opt} = \mathbf{Y} \cdot \mathbf{W}_c^{+}. \tag{5.12}$$

$\hat{\mathbf{X}}_{opt}$ is the input signal matrix with the smallest norm $\|\hat{\mathbf{X}}\|_F \le \|\mathbf{X}\|_F$ for each **X** such that $\mathbf{Y} = \mathbf{X} \cdot \mathbf{W}_c$. We could also state that $\hat{\mathbf{X}}_{opt}$ minimizes the error norm $\|\mathbf{X} \cdot \mathbf{W}_c - \mathbf{Y}\|_F$.

If, however, the elements of $\mathbf{W}_c$ are <u>complex</u>-valued, the **pseudoinverse** of $\mathbf{W}_c$ is calculated as

$$\mathbf{W}_c^{+} = \begin{cases} (\mathbf{W}_c^H \cdot \mathbf{W}_c)^{-1} \cdot \mathbf{W}_c^H, & M < N \\ \mathbf{W}_c^H \cdot (\mathbf{W}_c \cdot \mathbf{W}_c^H)^{-1}, & M > N \end{cases} \tag{5.13}$$

where $\mathbf{W}_c^H$ denotes the Hermetian (= complex conjugate elements combined with transposition) of matrix $\mathbf{W}_c$.

MIMO Channel Problem 2a is solved by means of a four-step approach. Under noise-free conditions ($\mathbf{N} = \mathbf{0}$), we wish to find the optimum channel estimates in the form of matrix $\hat{\mathbf{W}}_c = f_{2a}(\mathbf{X}, \mathbf{Y})$ such that the error norm $\|\mathbf{W}_c - \hat{\mathbf{W}}_c\|_F$ is minimized. The **channel identification algorithm** can be summarized as follows:

(1) Find an orthonormal basis $\mathbf{U}_x$ of the (known) input signal matrix $\mathbf{X}$.

(Use Gram-Schmidt procedure or take matrix $\mathbf{Q}$ obtained from *QR*-decomposition of $\mathbf{X}$. *Note*: Format of $\mathbf{X}$ is *K*-by-*N*. Size of $\mathbf{U}_x$ is *K*-by-*K*.)

(2) In the *K*-dimensional signal vector space spanned by $\mathbf{U}_x$, express the *N* column vectors of $\mathbf{X}$ by their projections onto the orthogonal axes of $\mathbf{U}_x$.

(Calculate the generalized Fourier coefficients of $\mathbf{X}$ with respect to $\mathbf{U}_x$ as

$$\mathbf{W}_x = \mathbf{U}_x \cdot \mathbf{X}. \tag{5.14}$$

Note: Observe that $\mathbf{U}_x \cdot \mathbf{U}_x^T = \mathbf{I}$ where $\mathbf{I}$ is a *K*-by-*K* identity matrix. Size of $\mathbf{W}_x$ is *K*-by-*N*.)

(3) In the *K*-dimensional signal vector space spanned by $\mathbf{U}_x$, express the *M* column vectors of output signal matrix $\mathbf{Y}$ by their projections onto the orthogonal axes of $\mathbf{U}_x$.

(Calculate the generalized Fourier coefficients of $\mathbf{Y}$ with respect to $\mathbf{U}_x$ as

$$\mathbf{W}_y = \mathbf{U}_x \cdot \mathbf{Y}. \tag{5.15}$$

Note: Size of $\mathbf{W}_y$ is *K*-by-*M*.)

(4) Calculate the pseudoinverse of $\mathbf{W}_x$ and find the optimum estimate of channel matrix $\mathbf{W}_c$ as

$$\hat{\mathbf{W}}_c = \mathbf{W}_x^+ \cdot \mathbf{W}_y. \tag{5.16}$$

(*Note*: Size of $\hat{\mathbf{W}}_c$ is *M*-by-*N*.)

For vast quantities of samples per column, *K*, and under real-time processing conditions, calculation of the inverse or pseudoinverse of $\mathbf{W}_c$ often requires prohibitively large numbers of arithmetic operations. To reduce the computational load, we may employ *QR*-decomposition of $\mathbf{W}_c$ prior to matrix inversion. If the factors $\mathbf{Q}$ and $\mathbf{R}$ of square matrix $\mathbf{W}_c$ are known, we may write (5.2) as

$$\mathbf{Y} = \mathbf{Q} \cdot \mathbf{R} \cdot \mathbf{X} \tag{5.17}$$

where $\mathbf{Q} \cdot \mathbf{Q}^T = \mathbf{I}$ and **R** is an upper-triangular matrix. Now, we substitute

$$\mathbf{A} = \mathbf{R} \cdot \mathbf{X}. \tag{5.18}$$

and left-multiply (5.17) by $\mathbf{Q}^{-1} = \mathbf{Q}^T$ to obtain the auxiliary matrix **A** as

$$\mathbf{A} = \mathbf{R} \cdot \mathbf{X} = \mathbf{Q}^T \cdot \mathbf{Y} \tag{5.19}$$

which can be done with $O(K^2)$ multiplications and additions. After that intermediate step, since **R** is a triangular matrix, the desired input signal matrix **X** is obtained by back substitution which requires an additional $O(K^2/2)$ multiplications and additions.

Two numerical examples shall now demonstrate these steps. We begin with a problem where all matrices are of square format The reader should have a quick look at the source code of MATLAB program "**ex5_1.m**" before running it and eventually changing the matrix entries.

EXAMPLE 5-1: Suppose we observe $K = 4$ time samples per input and per output port, respectively, of a (4×4) -channel transmission system. Since the system has $N = 4$ inputs and $M = 4$ outputs, the format of the unknown channel weight matrix $\mathbf{W}_c$ should be 4-by-4. Let us assume that all signals are real-valued and that the system inputs and outputs are observed as signal matrices

$$\mathbf{X} = \begin{pmatrix} 1 & -2 & 2 & 5 \\ 3 & 1 & 2 & 3 \\ 7 & 4 & 3 & -1 \\ 2 & 3 & 5 & 2 \end{pmatrix}, \quad \mathbf{Y} = \begin{pmatrix} 1 & 4 & 2 & 1 \\ 8 & -1 & 1 & 2 \\ 9 & 3 & 8 & 4 \\ 2 & 7 & -3 & 5 \end{pmatrix}. \tag{5.20}$$

(*Step 1*) Application of the *QR*-factorization procedure on input matrix **X** yields a four-dimensional (column-oriented) orthonormal basis in the form of

$$\mathbf{U}_x = \begin{pmatrix} -0.1260 & 0.7866 & 0.5337 & 0.2838 \\ -0.3780 & 0.2052 & 0.0862 & -0.8987 \\ -0.8819 & -0.0342 & -0.3338 & 0.3311 \\ -0.2520 & -0.5814 & 0.7722 & 0.0473 \end{pmatrix}. \tag{5.21}$$

(*Step 2*) **X** can be written as a linear combination of the mutually orthogonal column vectors $\mathbf{u}_k$, $k = 1, 2, 3, 4$, of the orthonormal basis $\mathbf{U}_x$. The generalized Fourier coefficients of

X with respect to $\mathbf{U}_x$ are then calculated to be

$$\mathbf{W}_x = \mathbf{U}_x \cdot \mathbf{X} = \begin{pmatrix} 6.5375 & 4.0249 & 4.3414 & 1.7637 \\ -0.9563 & -1.3901 & -4.5803 & -3.1578 \\ -2.6590 & 1.3876 & -1.1782 & -3.5162 \\ 3.5037 & 3.1532 & 0.8863 & -3.6816 \end{pmatrix}. \tag{5.22}$$

(Step 3) Next, we need the generalized Fourier coefficients of **Y** with respect to $\mathbf{U}_x$. These coefficients are determined by the projections of the column vectors of **Y** onto basis $\mathbf{U}_x$. We find them as

$$\mathbf{W}_y = \mathbf{U}_x \cdot \mathbf{Y} = \begin{pmatrix} 11.5379 & 2.2972 & 3.9531 & 5.0011 \\ 0.2421 & -7.7491 & 2.8349 & -4.1161 \\ -3.4977 & -2.1773 & -5.4618 & -0.6301 \\ 2.1411 & 2.2211 & 4.9502 & 1.9105 \end{pmatrix}. \tag{5.23}$$

(Step 4) Finally, since all matrices involved are of square size (4-by-4), we may avoid calculation of the pseudoinverse. Use of the inverse of matrix $\mathbf{W}_x$ in (5.22) then yields the channel matrix as

$$\hat{\mathbf{W}}_c = \mathbf{W}_c = \mathbf{W}_x^{-1} \cdot \mathbf{W}_y = \begin{pmatrix} 1.0437 & 0.1938 & 2.1875 & 0.0563 \\ 3.3250 & -3.2750 & -2.7500 & 0.2750 \\ -3.0250 & 4.1750 & 0.7500 & 0.8250 \\ 2.5312 & -2.2187 & -1.4375 & -0.0312 \end{pmatrix}. \tag{5.24}$$

Note that all matrices involved are of full rank, and the linear equation system $\mathbf{Y} = \mathbf{X} \cdot \mathbf{W}_c$ leads to a unique solution for the unknown matrix $\mathbf{W}_c$ with zero error norm, i.e., we have achieved the ultimate goal of

$$\left\| \mathbf{W}_c - \hat{\mathbf{W}}_c \right\|_F = 0. \tag{5.25}$$

▶◀

Whereas the above example lead to an ideal solution of MIMO Channel Problem 2a in a straightforward manner, the next example will demonstrate the subtleties of working with *non-square* matrix arrangements. Again, the executable source code is available as MATLAB program **"ex5_2.m"**.

EXAMPLE 5-2: As in the previous example, we observe $K = 4$ time samples per input and per output port, respectively. The system, however, shall now have $N = 2$ inputs and $M = 3$ outputs. Hence, the size of the unknown channel weight matrix will be 2-by-3. The linear system of equations, compactly written as $\mathbf{Y} = \mathbf{X} \cdot \mathbf{W}_c$, is inconsistent. We are now faced with the task of identifying an over-determined linear system with more equations than unknowns. The solution, in a minimum least squared error (LSE) sense, can be typically found as follows:

Suppose the input/output signal scenario is given as

$$\mathbf{X} = \begin{pmatrix} 1 & -2 \\ 3 & 1 \\ 7 & 4 \\ 2 & 3 \end{pmatrix}, \quad \mathbf{Y} = \begin{pmatrix} 1 & 4 & 2 \\ 8 & -1 & 1 \\ 9 & 3 & 8 \\ 2 & 7 & -3 \end{pmatrix} \tag{5.26}$$

which is a mere reduction of (5.20) from four to two and three columns in **X** and **Y**, respectively.

To find the "best possible" approximation of $\mathbf{W}_c$, let us proceed in four steps, as we did in the previous example.

(*Step 1*) By *QR*-factorization of **X** we find a new orthonormal basis as

$$\mathbf{U}_x = \begin{pmatrix} -0.1260 & 0.7866 & -0.1667 & 0.5811 \\ -0.3780 & 0.2052 & -0.7062 & -0.5624 \\ -0.8819 & -0.0342 & 0.4701 & -0.0101 \\ -0.2520 & -0.5814 & -0.5025 & 0.5882 \end{pmatrix} \tag{5.27}$$

whose two rightmost columns are different from those in (5.21). The first two columns are identical with those obtained in the previous example. Note that the last two rows of upper-triangular matrix **R**,

$$\mathbf{R} = \begin{pmatrix} -7.9373 & -4.4096 \\ 0 & -3.2489 \\ 0 & 0 \\ 0 & 0 \end{pmatrix}, \tag{5.28}$$

are both filled with zeros. The rank of our system has decreased from four (= full rank) to two.

(*Step 2*) Nevertheless, it is still possible to express **X** as a linear combination of the mutually orthogonal column vectors $\mathbf{u}_k$, $k = 1, 2, 3, 4$, of the (nonideal!) orthonormal basis $\mathbf{U}_x$. We obtain the 4-by-2 matrix of generalized Fourier coefficients of **X** with respect to $\mathbf{U}_x$ as

$$\mathbf{W}_x = \mathbf{U}_x \cdot \mathbf{X} = \begin{pmatrix} 2.2289 & 2.1148 \\ -5.8308 & -3.5510 \\ 2.2857 & 3.5797 \\ -4.3369 & -0.3226 \end{pmatrix}. \tag{5.29}$$

(*Step 3*) In a similar manner, we compute the projections of the $M = 3$ column vectors of **Y** onto basis $\mathbf{U}_x$ as

$$\mathbf{W}_y = \mathbf{U}_x \cdot \mathbf{Y} = \begin{pmatrix} 5.8283 & 2.2767 & -2.5423 \\ -6.2173 & -7.7724 & -4.5136 \\ 3.0549 & -2.1537 & 1.9926 \\ -8.2488 & 2.1837 & -6.8698 \end{pmatrix}. \tag{5.30}$$

(Step 4) We need both the pseudoinverse of $\mathbf{W}_x$ and the generalized Fourier coefficients of output matrix $\mathbf{Y}$ to calculate an optimum estimate of the unknown channel matrix. The desired weights of channel matrix $\mathbf{W}_c$ are then approximated (with minimum error norm!) by the matrix product

$$\hat{\mathbf{W}}_c = \mathbf{W}_x^+ \cdot \mathbf{W}_y = \begin{pmatrix} 1.6241 & 0.3609 & 1.4286 \\ -0.2947 & 0.3789 & -1.0000 \end{pmatrix}. \tag{5.31}$$

We should include an additional step to check the quality of our approximation. A reasonable approach is to assume $\hat{\mathbf{W}}_c$ and calculate $\hat{\mathbf{Y}}$ as

$$\hat{\mathbf{Y}} = \mathbf{X} \cdot \hat{\mathbf{W}}_c = \begin{pmatrix} 2.2135 & -0.3970 & 3.4286 \\ 4.5774 & 1.4617 & 3.2857 \\ 10.1895 & 4.0421 & 6.0000 \\ 2.3639 & 1.8586 & -0.1429 \end{pmatrix}. \tag{5.32}$$

It is readily seen that $\hat{\mathbf{Y}}$ very poorly approximates the (known) output signal matrix $\mathbf{Y}$. As we don't have available the "exact" channel matrix $\mathbf{W}_c$, the estimation quality can only be determined indirectly by the output error norm

$$\left\| \mathbf{Y} - \hat{\mathbf{Y}} \right\|_{\mathrm{F}} = 9.3314. \tag{5.33}$$

▸◂

5.3 FINITE SETS OF SIGNAL MATRICES

Finally, before we proceed with noisy channel scenarios, let us briefly investigate situations where the possible channel inputs are restricted to a finite set $\mathfrak{X}$ of, say P, signal matrices

$$\mathfrak{X} \in \{\mathbf{X}_1, \mathbf{X}_2, \ldots, \mathbf{X}_P\}. \tag{5.34}$$

Important applications include modern communication schemes where, by means of an extra **space-time coding** (STC) step, all columns of the pth input matrix $\mathbf{X}_p$, $p = 1, 2, ..., P$, are mutually orthogonal or even orthonormal vectors.

Moreover, we shall assume that all these "allowed" input matrices, or K-by-N frames of transmitted data, are known beforehand (*a-priori*) and stored in the memory of a receiver. That receiver shall have simultaneous access to each one of the M channel outputs. It can, therefore, employ the pth channel output signal matrix, taken from a finite set

$$\mathcal{Y} \in \{\mathbf{Y}_1, \mathbf{Y}_2, ..., \mathbf{Y}_P\}, \tag{5.35}$$

to recover the correct input signal matrix $\mathbf{X}_p$ by making a decision about the most likely frame number p out of P. If the receiver decides in favor of an estimate $\hat{p} = p$, a full K-by-N block of correctly detected data becomes available. On the other hand, that same data block is at least partially lost if the wrong decision ($\hat{p} \neq p$) is being made. As we shall see later on, the basic principle of the above decision method can also be successfully employed in receivers operating at the outputs of noisy MIMO channels.

In summary, we are now working on a special (discrete) type of *MIMO Channel Problem 1a* with a finite set of possible input matrices and ask the following questions:

- Knowing the true MIMO channel characteristics $\mathbf{W}_c$, is it always possible to recover the original messages $\mathbf{X}$ from noise-free output observations $\mathbf{Y}$?
- Moreover, are our estimates $\hat{\mathbf{X}}$ perfect (i.e., $\hat{\mathbf{X}} = \mathbf{X}$), or should we expect nonzero error norms?
- Under ideal conditions, are the solutions unique or just taken out of a set of P possible input messages?

The following example shall demonstrate a simple decision algorithm and possible errors caused by reduced-rank signal matrices. (See MATLAB program "**ex5_3.m**" for the source code used.) Note that we shall assume a noise-free MIMO channel model with a set of *a-priori* known input signal matrices and (pseudo)invertible channel matrices.

EXAMPLE 5-3: Let us observe four ($K = 4$) consecutive samples at each one of the three ($M = 3$) output ports of a MIMO channel. The measured output signal matrix is given by

$$\mathbf{Y} = \begin{pmatrix} 1 & 4 & 2 \\ 8 & -1 & 1 \\ 9 & 3 & 8 \\ 2 & 7 & -3 \end{pmatrix}. \tag{5.36}$$

The channel shall have two ($N = 2$) input ports and three ($M = 3$) outputs. We wish to find out which one of the two ($P = 2$) possible input signal matrices,

$$\mathbf{X}_1 = \begin{pmatrix} 4 & 3 \\ 7 & 2 \\ 2 & -3 \\ 1 & 5 \end{pmatrix}, \quad \mathbf{X}_2 = \begin{pmatrix} 1 & 2 \\ 3 & -1 \\ 7 & 1 \\ 2 & 6 \end{pmatrix} \tag{5.37}$$

has been transmitted over the channel. Once, based upon output observations $\mathbf{Y}$, we have made the decision about the correct index $p \in \{1, 2\}$, we should be able to recover the total amount of data contained in either $\mathbf{X}_1$ or $\mathbf{X}_2$. Since the set of possible input signal matrices, $\mathfrak{X} \in \{\mathbf{X}_1, \mathbf{X}_2\}$, is known *a-priori*, we may calculate in advance two orthonormal bases, $\mathbf{U}_{\mathbf{X}_1}$ and $\mathbf{U}_{\mathbf{X}_2}$, corresponding to $\mathbf{X}_1$ and $\mathbf{X}_2$, respectively. Using MATLAB function *qr*(...) for *QR*-decompositions of $\mathbf{X}_1$ and $\mathbf{X}_2$, we find these bases as

$$\mathbf{U}_{\mathbf{X}_1} = \begin{pmatrix} -0.478 & -0.478 & 0.3338 & -0.7714 \\ -0.8367 & -0.0810 & -0.4542 & 0.2952 \\ -0.2390 & -0.6020 & 0.7164 & 0.2594 \\ -0.1195 & 0.7525 & 0.4112 & 0.5004 \end{pmatrix}, \tag{5.38}$$

$$\mathbf{U}_{\mathbf{X}_2} = \begin{pmatrix} -0.1260 & 0.2824 & -0.7000 & -0.6437 \\ -0.3780 & -0.3059 & -0.6209 & 0.6149 \\ -0.8819 & -0.1647 & 0.3454 & -0.2753 \\ -0.2520 & 0.8942 & 0.0723 & 0.3629 \end{pmatrix}. \tag{5.39}$$

Input signal matrices $\mathbf{X}_1$ and $\mathbf{X}_2$ may be expressed in the form their Generalized Fourier Coefficients (GFC's) as

$$\mathbf{W}_{\mathbf{X}_1} = \mathbf{U}_{\mathbf{X}_1} \cdot \mathbf{X}_1 = \begin{pmatrix} -0.2333 & -5.7836 \\ -4.5271 & 0.1665 \\ -3.4778 & -2.7730 \\ 6.1120 & 2.4150 \end{pmatrix}, \tag{5.40}$$

$$\mathbf{W}_{\mathbf{X}_2} = \mathbf{U}_{\mathbf{X}_2} \cdot \mathbf{X}_2 = \begin{pmatrix} -5.4663 & -5.0968 \\ -4.4118 & 2.6185 \\ 0.4914 & -2.9053 \\ 3.6624 & 0.8518 \end{pmatrix}. \tag{5.41}$$

In a similar manner, we may utilize the two orthonormal bases to calculate the GFC's of the observed output signal matrix $\mathbf{Y}$ as

$$\mathbf{W}_{\mathbf{Y}_1} = \mathbf{U}_{\mathbf{X}_1} \cdot \mathbf{Y} = \begin{pmatrix} 3.0211 & -6.5655 & 4.2836 \\ -4.9822 & -2.5617 & -6.2735 \\ 1.9114 & 3.6109 & 3.8726 \\ 10.6018 & 3.5061 & 2.3016 \end{pmatrix}, \tag{5.42}$$

$$\mathbf{W}_{\mathbf{Y}_2} = \mathbf{U}_{\mathbf{X}_2} \cdot \mathbf{Y} = \begin{pmatrix} -5.4545 & -7.3925 & -3.6384 \\ -7.1831 & 1.2358 & -7.8734 \\ 0.3587 & -4.2535 & 1.6607 \\ 8.2779 & 0.8554 & -0.1203 \end{pmatrix}. \tag{5.43}$$

Then, an individual channel weighting matrix, $\mathbf{W}_{c,1}$ and $\mathbf{W}_{c,2}$, per orthonormal basis can be determined by right multiplication of

$$\mathbf{W}_{\mathbf{Y}_p} = \mathbf{W}_{c,p} \cdot \mathbf{W}_{\mathbf{X}_p}, \quad p = 1,2 \tag{5.44}$$

by the pseudoinverse of $\mathbf{W}_{\mathbf{X}_1}$ or $\mathbf{W}_{\mathbf{X}_2}$, respectively. Using the real-valued matrix elements of our example, we obtain the two approximations of the unknown (true) channel matrix $\mathbf{W}_c$ as

$$\mathbf{W}_{c,1} = \mathbf{W}_{\mathbf{X}_1}^{+} \cdot \mathbf{W}_{\mathbf{Y}_1} = \begin{pmatrix} 1.3921 & 0.0503 & 0.7846 \\ -0.6979 & 0.7392 & -1.0769 \end{pmatrix}, \tag{5.45}$$

$$\mathbf{W}_{c,2} = \mathbf{W}_{\mathbf{X}_2}^{+} \cdot \mathbf{W}_{\mathbf{Y}_2} = \begin{pmatrix} 1.5478 & 0.2326 & 1.0491 \\ -0.3062 & 1.1860 & -0.6163 \end{pmatrix}. \tag{5.46}$$

On the other hand, employing all the knowledge we have available, we may reversely estimate the transmitted input matrix $\mathbf{X}$ as either

$$\hat{\mathbf{X}}_1 = \mathbf{Y} \cdot \mathbf{W}_{c,1}^{+} = \begin{pmatrix} 2.9192 & 2.4162 \\ 5.2745 & 0.9054 \\ 7.7934 & 0.5419 \\ 5.8147 & 7.9132 \end{pmatrix} \tag{5.47}$$

or, in the vector space spanned by orthonormal basis $\mathbf{U}_{\mathbf{X}_2}$,

$$\hat{\mathbf{X}}_2 = \mathbf{Y} \cdot \mathbf{W}_{c,2}^{+} = \begin{pmatrix} 1.8973 & 2.5571 \\ 3.5601 & -0.6621 \\ 6.6751 & 0.8032 \\ 1.8484 & 5.9034 \end{pmatrix}. \tag{5.48}$$

Note that neither $\hat{\mathbf{X}}_1$ nor $\hat{\mathbf{X}}_2$ is completely identical to any of the two possible input signal matrices. Therefore, in the two (P = 2) four-dimensional (K = 4) vector spaces, we calculate and compare the spatial distances between all vectors involved. In other words, we need to find the two matrices whose column vectors are closest to each other. Taking the Frobenius error norms of matrix differences, a simple decision rule is obtained by calculating a **decision error matrix**, $\mathbf{\Gamma}$, in the form of

$$\mathbf{\Gamma} = \begin{pmatrix} \left\|\mathbf{X}_1 - \hat{\mathbf{X}}_1\right\|_F & \left\|\mathbf{X}_2 - \hat{\mathbf{X}}_1\right\|_F \\ \left\|\mathbf{X}_1 - \hat{\mathbf{X}}_2\right\|_F & \left\|\mathbf{X}_2 - \hat{\mathbf{X}}_2\right\|_F \end{pmatrix} = \begin{pmatrix} 9.1357 & 5.6314 \\ 7.8354 & 1.3115 \end{pmatrix}. \tag{5.49}$$

It is readily seen that the best possible decision $\hat{p}$ should be made according to the smallest diagonal element of matrix $\mathbf{\Gamma}$. Therefore, after searching for the minimum of all diagonal elements of $\mathbf{\Gamma}$, we pick

$$\hat{p} = \underset{i}{arg\,min}(\mathbf{\Gamma}_{i,i}) = 2, \quad i = 1, 2 \tag{5.50}$$

and decide in favor of input signal matrix $\mathbf{X}_2$.

►◄

5.4 SPACE-TIME BLOCK CODES

Now, what if the input signal matrices are intentionally chosen as orthogonal bases, and how can these mathematical concepts be employed in digital communications? To give an example, in **space-time block coding** (STC), we first design a finite set X consisting of P input signal matrices

$$\mathsf{X} \in \{\mathbf{U}_{\mathbf{X}_1}, \mathbf{U}_{\mathbf{X}_2}, \ldots, \mathbf{U}_{\mathbf{X}_P}\} \tag{5.51}$$

and then utilize each one of these orthogonal matrices to transmit more efficiently blocks of K information bits over a given MIMO channel with N inputs and M outputs. In the context of wireless communications, we wish to improve a transmission system's **spectral efficiency** defined as

$$\eta = \frac{r_b}{B} \tag{5.52}$$

where r_b is the data rate of the information-carrying bits at the system input and B is the available frequency bandwidth of the transmission channel. (In the following, we shall assume that bandwidth B is available through each one of the NM MIMO subchannels.) The block diagram of a digital transmitter employing space-time coding and N antennas is shown in Figure 5-2. Suppose we want to utilize Q-ary digital modulation in each subchannel. Then, at equidistant time intervals T_b, the **baseband encoder** (BBC) accepts from the input data source (S) binary digits in the form of a bit stream $d_b(iT_b)$ where i is an integer number counting the ith data bit. Depending on the kind of digital modulation scheme chosen, the baseband encoder uniquely assigns $log_2(Q)$ data bits to one out of the Q complex-valued, discrete-time baseband signals represented by

$$\underline{d}_s(kT_s) = a(kT_s) + jb(kT_s) \tag{5.53}$$

where T_s is the time interval between two successive symbols and k is the integer number counting the kth symbol.

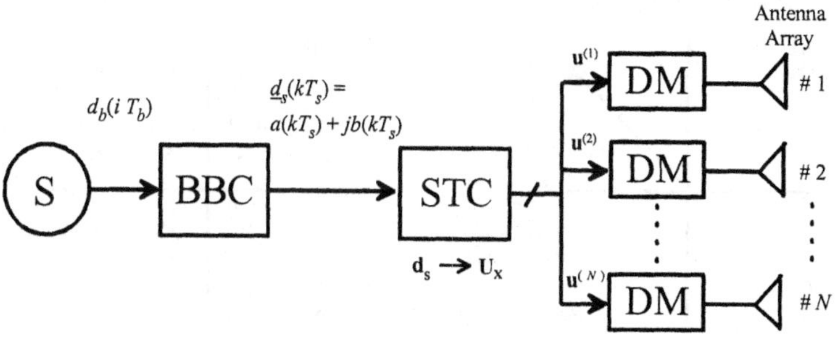

Figure 5-2. Block diagram of N-antenna wireless transmitter for generation of space-time encoded Q-ary digitally modulated signals. (S = binary information source; BBC = baseband symbol encoder; STC = space-time encoder; DM = Q-ary digital modulator).

Since every symbol represents $log_2(Q)$ information bits, the **symbol duration** T_s is given by

$$T_s = \frac{1}{r_s} = T_b \, log_2(Q) = \frac{log_2(Q)}{r_b} \tag{5.54}$$

where $r_s = 1/T_s$ is the **symbol rate**.

It is customary to display complex-valued baseband symbols $\underline{d}_s$ as vectors in two-dimensional signal space. Their real parts $a(kT_s)$ are usually called **inphase** (I) components. On the other hand, imaginary parts $b(kT_s)$ are frequently termed **quadrature phase** (Q) components. Two well-known examples of baseband encoding of binary digits into equiprobable complex-valued symbols are shown in Figure 5-3. The scheme on the left (*a*) is called **4-PSK** (PSK = phase-shift keying). Modulation format (*b*) on the righthand side maps a block of $log_2(16) = 4$ information bits to a complex-valued symbol vector. As both the carrier amplitude and the carrier phase are discretely modulated, the scheme may be categorized as 16-ary **quadrature amplitude modulation** (16-QAM). An abundance of other modulation formats exist - each one with its own merits and disadvantages in terms of bandwidth efficiency, noise sensitivity, etc. Design details and performance analyses of almost any relevant digital modulation scheme can be found in several excellent texts on digital communications such as the books by Proakis [1], Lee and Messerschmitt [2], Couch [3], to name just a few.

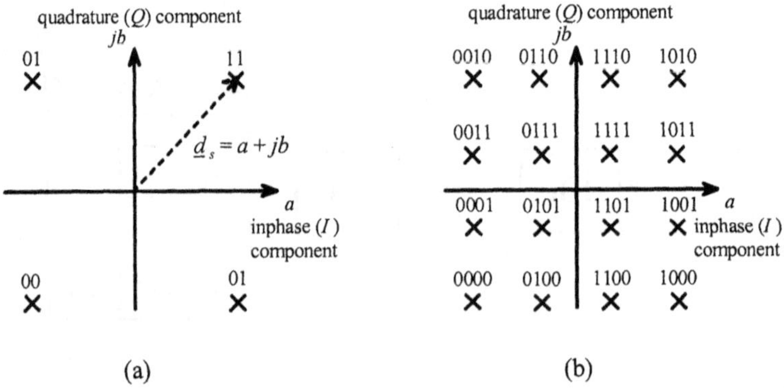

Figure 5-3. Examples of bit encoding rules for digitally modulated signals represented as vectors $\underline{d}_s(kT_s) = a(kT_s) + jb(kT_s)$ in complex signal-space at discrete sample times kT_s. (*a*) Four ($Q = 4$) phase-shift keying, 4 PSK; (*b*) Rectangular constellation for sixteen ($Q = 16$) quadrature amplitude modulation scheme, 16 QAM.

Then, starting at discrete time kT_s, the space-time encoder (STC) shown in the center part of Figure 5-2 stores a block of K complex-valued symbols, say $\underline{d}_s(k)$, $\underline{d}_s(k+1)$, ... $\underline{d}_s(k+K)$ where the symbol duration T_s is omitted for the sake of brevity. In vector notation, we write a block of symbols as

$$\mathbf{d}_s = (\underline{d}_s(k), \underline{d}_s(k+1), \ldots, \underline{d}_s(k+K))^T . \tag{5.55}$$

Using the elements of $\mathbf{d}_s$ and assuming a given number N of transmit antennas, the space-time encoder then generates a code-specific, scaled unitary **transmission matrix** $\mathbf{U}_X$ whose N columns are orthogonal to each other. The nth column of $\mathbf{U}_X$ is denoted by $\mathbf{u}^{(n)}$. Its ℓ entries represent the complex symbols serving as inputs to the nth digital modulator (DM). Let ℓ be the number of symbol periods (i.e., total time = ℓT_s) required to transmit a space-time coded symbol block through the nth modulator and antenna path. A space-time encoder with K input symbols and ℓ transmit periods may then be characterized by its **space-time code rate** (r_{STC}) given by the ratio

$$r_{STC} = \frac{K}{\ell} . \tag{5.56}$$

A number of K complex symbols is transmitted over ℓ time intervals. Ideally, a perfect space-time code is one with full rate (i.e., r_{STC} = 1) where no extra transmission time periods are required. Another important quantity is readily derived if we substitute r_{STC} into equations (5.52) and (5.54), respectively. For a Q-ary digital modulation scheme, we may define the **spectral efficiency** (η_{STC}) of a space-time block code with code rate r_{STC} as

$$\eta_{STC} = \frac{r_b}{B} = \frac{r_s \, log_2(Q) r_{STC}}{r_s} = \frac{K \, log_2(Q)}{\ell} . \tag{5.57}$$

5.5 THE ALAMOUTI SCHEME

Original and fundamental ideas on space-time coding and optimal code construction came from Tarokh, Seshadri, and Calderbank in their March 1998 paper [4]. Later on that same year, in October of 1998, Siavash Alamouti proposed a simple, yet fundamentally new space-time coding method with N = 2 transmit antennas and, as a minimum, M = 1 receive antenna [5]. This work, now known as the Alamouti scheme, also triggered a large number of theoretical and practical investigations into the research areas of space-time coding, optimal transmit matrices, rate and performance analyses, etc. Indepth treatments of space-time codes, including the Alamouti scheme, have recently been published in a few modern textbooks on these subjects. Recommended texts include the books by Vucetic and Yuan [6], Paulraj *et al.* [7], Larsson and Stoica [8], and Hottinen *et al.* [9].

Before we go through a practical example, let us briefly summarize the advantages of the **Alamouti scheme**. According to Alamouti's original paper [5], the new scheme has the following properties that distinguish it from conventional methods:

* Improves the signal quality at the receiver by simple processing across two transmit antennas.

* No bandwidth expansion, as redundancy is applied in space across multiple antennas, not in time or frequency.

* Works without any feedback from the receiver to the transmitter.

* Has small computation complexity - as compared with existing approaches.

* Offers decreased sensitivity to signal fading on the channel and, therefore, allows for utilization of higher level modulation schemes to increase the effective data rate.

An example with arbitrarily chosen input data shall now demonstrate the steps necessary to implement Alamouti's transmit diversity technique. MATLAB program "**ex5_4.m**" should be run to study details of computational steps.

EXAMPLE 5-4: We consider a MISO system with two ($N = 2$) transmit antennas and, as the abbreviation "MISO" implies, a single receive antenna ($M = 1$). Suppose we wish to transmit, at fixed bit rate r_b, a short message consisting of eight information bits. The input data stream is represented by column vector

$$\mathbf{d}_b = (0\,0\,1\,0\,1\,1\,0\,1)^T \tag{5.58}$$

The 4-PSK constellation scheme of Figure 5-3 (*a*) shall be used to encode $Q = 4$ different phase states of the RF carrier signal. According to (5.51), the symbol rate r_s is given by

$$r_s = \frac{r_b}{log_2(4)} = \frac{r_b}{2}\,. \tag{5.59}$$

We shall employ complex-valued baseband symbols $\underline{d}_s$ of unit magnitude (= vector length in the constellation diagram). Thus, after consecutively encoding blocks of two bits ("dibits") at integer multiples of symbol time T_s, the **frame of baseband symbols** ($\mathbf{d}_s$) can be written as a column vector of length L_F. Here, the frame consists of $L_F = 4$ symbols and $\mathbf{d}_s$ is given by

$$\mathbf{d}_s = (\underline{d}_1, \underline{d}_2, \underline{d}_3, \underline{d}_4)^T = \frac{1}{2}\sqrt{2}\big((-1-j), (1-j), (1+j), (-1+j)\big)^T . \tag{5.60}$$

Note that the 4-PSK symbol coding scheme may be changed but should be known at the receiver. For the Alamouti scheme, we have $N = \ell = 2$ and, therefore, the scheme achieves the full space-time code rate of $r_{STC} = 1$. Next, space-time encoding of symbol frame $\mathbf{d}_s$ is performed by taking $\ell = 2$ successive blocks of two ($N = 2$) symbols and generating **scaled-orthogonal code matrices**

$$\mathbf{U}_{\mathbf{X}_1} = \begin{pmatrix} \underline{d}_1 & \underline{d}_2 \\ -\underline{d}_2^* & \underline{d}_1^* \end{pmatrix} = \frac{1}{2}\sqrt{2}\begin{pmatrix} -1-j & 1-j \\ -1-j & -1+j \end{pmatrix} = (\mathbf{u}_{\mathbf{X}_1}^{(1)}, \mathbf{u}_{\mathbf{X}_1}^{(2)}), \tag{5.61}$$

$$\mathbf{U}_{\mathbf{X}_2} = \begin{pmatrix} \underline{d}_3 & \underline{d}_4 \\ -\underline{d}_4^* & \underline{d}_3^* \end{pmatrix} = \frac{1}{2}\sqrt{2}\begin{pmatrix} 1+j & -1+j \\ 1+j & 1-j \end{pmatrix} = (\mathbf{u}_{\mathbf{X}_2}^{(1)}, \mathbf{u}_{\mathbf{X}_2}^{(2)}) \tag{5.62}$$

where superscript $^{(1)}$ and $^{(2)}$, respectively, denotes the number of the nth digital modulator/antenna path. In our example, we have intentionally restricted the frame length to $L_F = 4$ and, hence, obtain no more than $L_F/2 = 2$ code matrices. During the first ($k = 1$) symbol period T_s, symbol $\underline{d}_1$ is transmitted via antenna #1, and $\underline{d}_2$ is simultaneously transmitted by antenna #2. According to code matrix $\mathbf{U}_{\mathbf{X}_1}$, during the next symbol period ($k = 2$), the negative, complex conjugate of symbol $\underline{d}_2$ is output by antenna #1 while the complex conjugate of symbol $\underline{d}_1$ goes through antenna #2. Similarly, as required by code matrix $\mathbf{U}_{\mathbf{X}_2}$, $\underline{d}_3$ and $\underline{d}_4$ are the symbols to be transmitted via antennas #1 and #2, respectively, in the next time interval ($k = 3$). Finally, during the fourth ($k = 4$) symbol interval, the negative and complex conjugate version of symbol $\underline{d}_4$ is output by antenna #1, and $\underline{d}_3^*$ is simultaneously transmitted by antenna #2. This completes our short sequence of parallel transmissions over $K = 4$ symbol periods. The space-time coding scheme proposed by Alamouti uses mutually orthogonal column vectors $\mathbf{u}^{(1)}$ and $\mathbf{u}^{(2)}$ in each code matrix. This is easily verified for $\mathbf{U}_{\mathbf{X}_1}$ (or any other code matrix in the frame) by showing that the inner product of the $N = 2$ columns is zero. Here, we have

$$\left\langle \mathbf{u}_{\mathbf{X}_1}^{(1)}, \mathbf{u}_{\mathbf{X}_1}^{(2)} \right\rangle = (\underline{d}_1, -\underline{d}_2^*) \cdot (\underline{d}_2, \underline{d}_1^*)^H = \underline{d}_1 \cdot \underline{d}_2^* - \underline{d}_2^* \cdot \underline{d}_1 = 0 , \tag{5.63}$$

$$\left\langle \mathbf{u}_{\mathbf{X}_2}^{(1)}, \mathbf{u}_{\mathbf{X}_2}^{(2)} \right\rangle = (\underline{d}_3, -\underline{d}_4^*) \cdot (\underline{d}_4, \underline{d}_3^*)^H = \underline{d}_3 \cdot \underline{d}_4^* - \underline{d}_4^* \cdot \underline{d}_3 = 0 . \tag{5.64}$$

Moreover, we observe that **all code matrices constitute orthogonal bases**. In our case, this is verified by calculation of

$$\mathbf{U}_{\mathbf{X}_1} \cdot \mathbf{U}_{\mathbf{X}_1}^H = \frac{1}{2}\begin{pmatrix} -1-j & 1-j \\ -1-j & -1+j \end{pmatrix} \cdot \begin{pmatrix} -1+j & -1+j \\ 1+j & -1-j \end{pmatrix} = 2\begin{pmatrix} 1 & 0 \\ 0 & 1 \end{pmatrix}$$
$$= (|\underline{d}_1|^2 + |\underline{d}_2|^2) \cdot \begin{pmatrix} 1 & 0 \\ 0 & 1 \end{pmatrix} \quad (5.65)$$

$$\mathbf{U}_{\mathbf{X}_2} \cdot \mathbf{U}_{\mathbf{X}_2}^H = \frac{1}{2}\begin{pmatrix} 1+j & -1+j \\ 1+j & 1-j \end{pmatrix} \cdot \begin{pmatrix} 1-j & 1-j \\ -1-j & 1+j \end{pmatrix} = 2\begin{pmatrix} 1 & 0 \\ 0 & 1 \end{pmatrix}$$
$$= (|\underline{d}_3|^2 + |\underline{d}_4|^2) \cdot \begin{pmatrix} 1 & 0 \\ 0 & 1 \end{pmatrix} \quad (5.66)$$

A simplified model of the noisy MISO transmission channel and of the receiver is shown in Figure 5-4. We assume unit symbol weighting ($\underline{w}_{T1} = \underline{w}_{T2} = 1$) by the two transmit antennas (subscript T) and by the single receive (subscript R) antenna ($\underline{w}_{R1} = 1$), respectively. During the first space-time code block, the transmit symbols leaving antenna #1 at discrete symbol times $k = 1, 2$ are described as

$$\underline{x}_1^{(1)} = \underline{x}_1 = \underline{d}_1 \underline{w}_{T1} = \underline{d}_1 = \frac{1}{2}\sqrt{2}(-1-j). \quad (5.67)$$

$$\underline{x}_2^{(1)} = -\underline{x}_2^* = -\underline{d}_2^* \underline{w}_{T1} = -\underline{d}_2^* = \frac{1}{2}\sqrt{2}(-1-j). \quad (5.68)$$

In a similar manner, we denote the symbols emanating from antenna #2 by

$$\underline{x}_1^{(2)} = \underline{x}_2 = \underline{d}_2 \underline{w}_{T2} = \underline{d}_2 = \frac{1}{2}\sqrt{2}(1-j). \quad (5.69)$$

$$\underline{x}_2^{(2)} = \underline{x}_1^* = \underline{d}_1^* \underline{w}_{T1} = \underline{d}_1^* = \frac{1}{2}\sqrt{2}(-1+j). \quad (5.70)$$

There are two paths whose fading characteristics at any discrete sample time k are described by complex **channel coefficients** $\underline{h}_{n,m}(k)$ from the nth transmit antenna to the mth receive antenna in the form of

$$\underline{h}_{1,1}(k) = \underline{h}_1(k) = |\underline{h}_1(k)| exp(j\phi_1(\mathrm{k})), \quad (5.71)$$

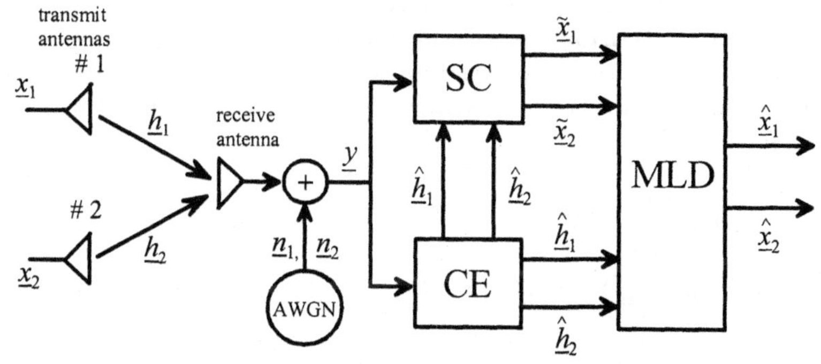

Figure 5-4. System model of QPSK receiver of Example 5-4. (AWGN = additive white Gaussian noise, CE = channel estimator; SC = signal combiner; MLD = maximum likelihood detector).

$$\underline{h}_{2,1}(k) = \underline{h}_2(k) = |\underline{h}_2(k)| exp(j\phi_2(\mathrm{k})). \tag{5.72}$$

We assume that the elements of vector $\mathbf{h} = (\underline{h}_1, \underline{h}_2)^T$ are constant over the **channel's coherence time** T_c. That is, for T_c greater than any two consecutive symbol time epochs, say numbers k and k+1, we have

$$\mathbf{h}(k) = \mathbf{h}(k+1). \tag{5.73}$$

Using the above notation, the two symbols received at discrete times $k = 1$ and $k = 2$ may be written as

$$\begin{aligned} \underline{y}(1) &= \underline{h}_1(1)\underline{x}_1 + \underline{h}_2(1)\underline{x}_2 + \underline{n}(1) \\ &= \underline{h}_1(1)\underline{d}_1 + \underline{h}_2(1)\underline{d}_2 + \underline{n}(1), \end{aligned} \tag{5.74}$$

$$\begin{aligned} \underline{y}(2) &= -\underline{h}_1(2)\underline{x}_2^* + \underline{h}_2(2)\underline{x}_1^* + \underline{n}(2) \\ &= -\underline{h}_1(2)\underline{d}_2^* + \underline{h}_2(2)\underline{d}_1^* + \underline{n}(2) \end{aligned} \tag{5.75}$$

where $\underline{n}(1)$ and $\underline{n}(2)$ are complex-valued **additive white Gaussian noise** (**AWGN**) samples with zero mean and power spectral density $N_0/2$ per dimension. As shown in Figure 5-4, we employ a **channel estimator** (**CE**) to derive the **channel state information** (**CSI**) in the form of channel vector $\hat{\mathbf{h}}$. As before, superscripts "hat" (^) are used to distinguish estimates from the true parameters. Note that the elements of vectors $\mathbf{h}$ and $\hat{\mathbf{h}}$ are constant over two consecutive symbol time intervals, i.e., $\underline{h}_1(1) = \underline{h}_1(2) = \underline{h}_1$ and $\underline{h}_2(1) = \underline{h}_2(2) = \underline{h}_2$, respectively. CSI updates are calculated after each space-time code block.

By means of a **signal combiner** (**SC**), we combine the received signals $\underline{y}(1)$ and $\underline{y}(2)$ with the channel state information represented by the estimates in vector $\hat{\mathbf{h}}$. In the SC block, **decision statistics** (identified by superscripts "tilde" (~)) are computed in the form of variables

$$\begin{aligned}\tilde{\underline{x}}(1) &= \hat{\underline{h}}_1^* \underline{y}(1) + \hat{\underline{h}}_2 \underline{y}^*(2)\\ &= (|\hat{\underline{h}}_1|^2 + |\hat{\underline{h}}_2|^2)\underline{x}_1 + \hat{\underline{h}}_1^* \underline{n}(1) + \hat{\underline{h}}_2 \underline{n}^*(2),\end{aligned} \tag{5.76}$$

$$\begin{aligned}\tilde{\underline{x}}(2) &= \hat{\underline{h}}_2^* \underline{y}(1) - \hat{\underline{h}}_1 \underline{y}^*(2)\\ &= (|\hat{\underline{h}}_1|^2 + |\hat{\underline{h}}_2|^2)\underline{x}_2 - \hat{\underline{h}}_1 \underline{n}^*(2) + \hat{\underline{h}}_2^* \underline{n}(1).\end{aligned} \tag{5.77}$$

These variables are only functions of $\underline{x}_1$ and $\underline{x}_2$, respectively. At the receive side, decisions about the transmit symbols $\underline{x}_1$ and $\underline{x}_2$ can therefore be made separately. It is convenient to introduce a channel matrix $\mathbf{H}$ and a scalar channel parameter α in such a way that

$$\mathbf{H} = \begin{pmatrix} \underline{h}_1 & \underline{h}_2 \\ \underline{h}_2^* & -\underline{h}_1^* \end{pmatrix}, \tag{5.78}$$

$$\begin{aligned}\mathbf{H}^H \cdot \mathbf{H} &= \begin{pmatrix} \underline{h}_1^* & \underline{h}_2 \\ \underline{h}_2^* & -\underline{h}_1 \end{pmatrix} \cdot \begin{pmatrix} \underline{h}_1 & \underline{h}_2 \\ \underline{h}_2^* & -\underline{h}_1^* \end{pmatrix}\\ &= (|\underline{h}_1|^2 + |\underline{h}_2|^2)\mathbf{I}_2 = \alpha \mathbf{I}_2\end{aligned} \tag{5.79}$$

where $\mathbf{I}_2$ is a two-by-two identity matrix.

Finally, a **maximum likelihood detector** (**MLD**) is employed to find the two signal points closest to the transmitted ones. Its task is to minimize the distance metric between vector $\tilde{\mathbf{x}}$ and all possible channel-scaled transmit vectors $\alpha\hat{\mathbf{x}}$. Mathematically, we formulate this task in vector form as

$$\hat{\mathbf{x}} = \underset{\hat{\mathbf{x}} \in \mathcal{X}}{arg\,min} \|\tilde{\mathbf{x}} - \alpha\hat{\mathbf{x}}\|_F^2 \tag{5.80}$$

where $\mathcal{X} = \{\underline{d}_1, \underline{d}_2, \underline{d}_3, \underline{d}_4\}$ is the set of all possible complex symbol vectors, and $\|\ldots\|_F^2$ is the squared Frobenius norm. Thus, we may decompose (5.80) into the following two ML decision rules:

$$\begin{aligned}\hat{\underline{x}}_1 &= \arg\min_{\hat{\underline{x}}_1 \in \mathrm{X}} \left|\tilde{\underline{x}}_1 - \alpha \hat{\underline{x}}_1\right|^2 \\ &= \arg\min_{\hat{\underline{x}}_1 \in \mathrm{X}} \left|\hat{\underline{h}}_1^* \underline{y}(1) + \hat{\underline{h}}_2 \underline{y}^*(2) - \alpha \hat{\underline{x}}_1\right|^2\end{aligned} \tag{5.81}$$

$$\begin{aligned}\hat{\underline{x}}_2 &= \arg\min_{\hat{\underline{x}}_2 \in \mathrm{X}} \left|\tilde{\underline{x}}_2 - \alpha \hat{\underline{x}}_2\right|^2 \\ &= \arg\min_{\hat{\underline{x}}_2 \in \mathrm{X}} \left|\hat{\underline{h}}_2^* \underline{y}(1) - \hat{\underline{h}}_1 \underline{y}^*(2) - \alpha \hat{\underline{x}}_2\right|^2\end{aligned} \tag{5.82}$$

In MATLAB program "**ex5_4.m**", constant channel coefficients are arbitrarily chosen as

$$\underline{h}_1 = 0.1\, exp(j\frac{\pi}{4}) = 0.0707(1+j)\,, \tag{5.83}$$

$$\underline{h}_2 = 0.3\, exp(j\frac{\pi}{3}) = 0.1500 + j0.2598\,. \tag{5.84}$$

Complex noise samples are also arbitrarily selected as $\underline{n}_1 = 0.0010 + j0.0020$ and $\underline{n}_2 = -0.0030 + j0.0030$. Calculation of all squared magnitudes according to (5.81) and (5.82) yields two separate error vectors

$$\xi_1 = (0.0000, 0.0197, 0.0402, 0.0205)^T\,, \tag{5.85}$$

$$\xi_2 = (0.0206, 0.0000, 0.0205, 0.0411)^T\,, \tag{5.86}$$

As the first element of vector ξ_1 represents that vector's minimum, the first decision ($\hat{\underline{x}}_1$) should be made in favor of $\underline{d}_1$. In a similar manner, we select $\hat{\underline{x}}_2 = \underline{d}_2$ as the ML estimate of the second symbol.

By repeating the above decision processes for code matrix $\mathbf{U}_{\mathbf{X}_2}$ with the same channel and noise parameters, we find new estimates $\hat{\underline{x}}_1 = \underline{d}_3$ and $\hat{\underline{x}}_2 = \underline{d}_4$, respectively, for the next block of input symbols. Note that zero symbol errors occurred and, hence, it is possible to perfectly recover all of the eight bits in the original datastream $\mathbf{d}_b$.

►◄

A fully featured Alamouti space-time codec with two transmit antennas and one receive antenna is available on the CD-ROM under the program name of "**Alamouti.m**". In that simulation program complex-valued AWGN samples are needed. These noise samples are generated by means of MATLAB function "**gnoise.m**" with the two control parameters *sig* (standard deviation of Gausian noise process) and *leng* (= number of samples in noise vector). By variation of the number of data frame's QPSK input symbols, of the noise level, and of the complex channel coefficients, program "**Alamouti.m**" can be used to simulate various transmission scenarios. It calculates the number of symbol errors per block and the symbol error rate for sets of arbitrarily chosen noise parameters and channel coefficients. It is rather straight-forward to modify the program to incorporate dynamic channel characteristics, varying noise levels, and other code matrices. Currently, various researchers concentrate their efforts on finding optimum scaled-orthogonal space-time code matrices. State-of-the-art STC designs at the time of completion of this manuscript are available in references [10] and [11]. The reader is encouraged to carry out simulation experiments with the space-time code matrices listed in Appendix C.

5.6 CHAPTER 5 PROBLEMS

5.1 Suppose we wish to generate code matrices in the form of orthonormal bases $\mathbf{U_X}$. For a 4-PSK signal constellation and an Alamouti block coding scheme with N = 2 transmit antennas and a single receive antenna (i.e., M = 1), what would be the length (= magnitude) of all complex signal vectors $\underline{d}_s(k)$ in the 4-PSK scheme.

5.2 Verify the expressions for the decision statistics ($\underline{\tilde{x}}_1$ and $\underline{\tilde{x}}_2$) in the second lines of equations (5.76) and (5.77) by substitution of the received signals $\underline{y}(1)$ and $\underline{y}(2)$ from equations (5.74) and (5.75), respectively.

5.3 Design a receiver for the scaled-orthogonal space-time block code $\mathbf{X}_3^{(c)}$ given by equation C-6 in Appendix C. Draw a receiver block diagram and develop a MATLAB program, say "**STC_X3.m**", to test the STC codec with 4-PSK input symbols and a fixed set of complex channel coefficients. *Hint*: As a starting point, try to modify MATLAB program "**Alamouti.m**".

REFERENCES

[1] J. G. Proakis. *Digital Communications*. McGraw-Hill, New York, N.Y., 4th edition, 2000.

[2] E. A. Lee and D. G. Messerschmitt. *Digital Communication*. Kluwer Academic Publishers, Boston, MA, 2nd edition, 1993.

[3] L. W. Couch. *Digital and Analog Communication Systems*. Prentice Hall, Inc., Englewood Cliffs, N. J., 6th edition, 2000.

[4] V. Tarokh, N. Seshadri, and A. R. Calderbank. "Space-time Codes for High Data Rate Wireless Communication: Performance Criterion and Code Construction," *IEEE Transactions on Information Theory*, Vol. 44, No. 2, March 1998, pp. 744 - 765.

[5] S. M. Alamouti. "A Simple Transmit Diversity Technique for Wireless Communications," *IEEE Journal on Selected Areas in Communications,* Vol. 16, No. 8, October 1998, pp. 1451 - 1458.

[6] B. Vucetic and J. Yuan. *Space-Time Coding*. John Wiley & Sons, Inc., Chichester, West Sussex, England, 2003.

[7] A. Paulraj, R. Nabar, and D. Gore. *Introduction to Space-Time Wireless Communications*. Cambridge University Press, Cambridge, United Kingdom, 2003.

[8] E. G. Larsson and P. Stoica. *Space-Time Block Coding for Wireless Communications.* Cambridge University Press, Cambridge, United Kingdom, 2003.

[9] A. Hottinen, O. Tirkkonen, and R. Wichman. *Multi-antenna Transceiver Techniques for 3G and Beyond.* John Wiley & Sons, Inc., Chichester, West Sussex, England, 2003.

[10] W. Zhao, G. Leus, and G. B. Giannakis. "Orthogonal Design of Unitary Constellations and Trellis-Coded Noncoherent Space-Time Systems," *IEEE Transactions on Information Theory*, Vol. 50, No. 6, June 2004, pp. 1319 - 1327.

[11] H.-F. Lu, P. V. Kumar, and H. Chung. "On Orthogonal Designs and Space-Time Codes," *IEEE Communications Letters*, Vol. 8, No. 4, April 2004, pp. 220 - 222.

APPENDIX A

VECTOR OPERATIONS

Suppose **a**, **b**, **c**, **d** are column vectors in a coordinate system with column unit vectors $\mathbf{u}_1$, $\mathbf{u}_2$, and $\mathbf{u}_3$.

We write, e.g., vector **a** in the form of $\mathbf{a} = a_1\mathbf{u}_1 + a_2\mathbf{u}_2 + a_3\mathbf{u}_3$ with vector components a_1, a_2, a_3.

Variables α and β are scalars.

Then the following laws and rules apply.

Commutative law for addition:

$$\mathbf{a} + \mathbf{b} = \mathbf{b} + \mathbf{a} \tag{A-1}$$

Associative law for addition:

$$\mathbf{a} + \mathbf{b} + \mathbf{c} = (\mathbf{a} + \mathbf{b}) + \mathbf{c} = \mathbf{a} + (\mathbf{b} + \mathbf{c}) \tag{A-2}$$

Associative law for scalar multiplication:

$$\alpha\beta\mathbf{a} = \alpha(\beta\mathbf{a}) = (\alpha\beta)\mathbf{a} = \beta(\alpha\mathbf{a}) \tag{A-3}$$

Distributive laws:

$$\alpha(\mathbf{a} + \mathbf{b}) = \alpha\mathbf{a} + \alpha\mathbf{b} \tag{A-4}$$

$$(\alpha + \beta)\mathbf{a} = \alpha\mathbf{a} + \beta\mathbf{a} \tag{A-5}$$

Dot (= scalar) product:

$$\mathbf{a} \cdot \mathbf{b} = |\mathbf{a}||\mathbf{b}| \cos(\angle \mathbf{a},\mathbf{b}),\ 0 \le \angle \mathbf{a},\mathbf{b} \le \pi \text{ (=angle between } \mathbf{a} \text{ and } \mathbf{b}) \qquad \text{(A-6)}$$

$$\mathbf{a} \cdot \mathbf{b} = \mathbf{b} \cdot \mathbf{a} \qquad \text{(A-7)}$$

$$\mathbf{a} \cdot (\mathbf{b} + \mathbf{c}) = \mathbf{a} \cdot \mathbf{b} + \mathbf{a} \cdot \mathbf{c} \qquad \text{(A-8)}$$

$$\mathbf{a} \cdot \mathbf{b} = a_1 b_1 + a_2 b_2 + a_3 b_3 \qquad \text{(A-9)}$$

Cross (= vector) product:

$$\mathbf{a} \times \mathbf{b} = |\mathbf{a}||\mathbf{b}| \sin(\angle \mathbf{a},\mathbf{b}) \mathbf{u}_\perp,$$

($\mathbf{u}_\perp$ = unit vector perpendicular to plane of **a** and **b**.
a, **b**, $\mathbf{u}_\perp$ form a right-handed system.) (A-10)

$$\mathbf{a} \times \mathbf{b} = \begin{vmatrix} \mathbf{u}_1 & \mathbf{u}_2 & \mathbf{u}_3 \\ a_1 & a_2 & a_3 \\ b_1 & b_2 & b_3 \end{vmatrix} \qquad \text{(A-11)}$$

$$= (a_2 b_3 - a_3 b_2)\mathbf{u}_1 + (a_3 b_1 - a_1 b_3)\mathbf{u}_2 + (a_1 b_2 - a_2 b_1)\mathbf{u}_3$$

$$\mathbf{a} \times \mathbf{b} = -\mathbf{b} \times \mathbf{a} \qquad \text{(A-12)}$$

$$\mathbf{a} \times (\mathbf{b} + \mathbf{c}) = \mathbf{a} \times \mathbf{b} + \mathbf{a} \times \mathbf{c} \qquad \text{(A-13)}$$

Mixed dot and vector products:

$$\mathbf{a} \times (\mathbf{b} \times \mathbf{c}) = \mathbf{b}(\mathbf{a} \cdot \mathbf{c}) - \mathbf{c}(\mathbf{a} \cdot \mathbf{b}) \qquad \text{(A-14)}$$

$$(\mathbf{a}\times\mathbf{b})\times\mathbf{c}=\mathbf{b}(\mathbf{a}\cdot\mathbf{c})-\mathbf{a}(\mathbf{b}\cdot\mathbf{c}) \tag{A-15}$$

$$\begin{aligned}(\mathbf{a}\times\mathbf{b})\times(\mathbf{c}\times\mathbf{d})&=\mathbf{c}(\mathbf{a}\cdot(\mathbf{b}\times\mathbf{d}))-\mathbf{d}(\mathbf{a}\cdot(\mathbf{b}\times\mathbf{c}))\\&=\mathbf{b}(\mathbf{a}\cdot(\mathbf{c}\times\mathbf{d}))-\mathbf{a}(\mathbf{b}\cdot(\mathbf{c}\times\mathbf{d}))\end{aligned} \tag{A-16}$$

$$\begin{aligned}\mathbf{a}\cdot(\mathbf{b}\times\mathbf{c})&=\begin{vmatrix}a_1 & a_2 & a_3\\ b_1 & b_2 & b_3\\ c_1 & c_2 & c_3\end{vmatrix}\\&=a_1(b_2c_3-b_3c_2)-a_2(b_1c_3-b_3c_1)+a_3(b_1c_2-b_2c_1)\end{aligned} \tag{A-17}$$

Derivatives of vectors $\mathbf{a}(\alpha)=a_1(\alpha)\mathbf{u}_1+a_2(\alpha)\mathbf{u}_2+a_3(\alpha)\mathbf{u}_3$, $\mathbf{b}(\alpha)$, $\mathbf{c}(\alpha)$, ...:

$$\frac{d\mathbf{a}}{d\alpha}=\frac{da_1}{d\alpha}\mathbf{u}_1+\frac{da_2}{d\alpha}\mathbf{u}_2+\frac{da_3}{d\alpha}\mathbf{u}_3 \tag{A-18}$$

$$\frac{d}{d\alpha}(\mathbf{a}\cdot\mathbf{b})=\mathbf{a}\cdot\frac{d\mathbf{b}}{d\alpha}+\frac{d\mathbf{a}}{d\alpha}\cdot\mathbf{b} \tag{A-19}$$

$$\frac{d}{d\alpha}(\mathbf{a}\times\mathbf{b})=\mathbf{a}\times\frac{d\mathbf{b}}{d\alpha}+\frac{d\mathbf{a}}{d\alpha}\times\mathbf{b} \tag{A-20}$$

$$\frac{d}{d\alpha}(\mathbf{a}\cdot(\mathbf{b}\times\mathbf{c}))=\frac{d\mathbf{a}}{d\alpha}\cdot(\mathbf{b}\times\mathbf{c})+\mathbf{a}\cdot(\frac{d\mathbf{b}}{d\alpha}\times\mathbf{c})+\mathbf{a}\cdot(\mathbf{b}\times\frac{d\mathbf{c}}{d\alpha}) \tag{A-21}$$

APPENDIX B

MATRIX ALGEBRA

A matrix **A** of order N by M is a rectangular array of NM quantities organized in N rows and M columns. If the number of rows is equal to the number of columns ($N = M$), **A** is called a square matrix. $a_{n,m}$ is the (n,m)th element of **A** where $n = 1, 2, ..., N$ denotes the nth row and $m = 1, 2, ..., M$ is the mth column.

If **A** has a single row ($N = 1$), **A** is a row vector.

If **A** has a single column ($M = 1$), **A** is a column vector.

Interchanging of rows and columns in **A** yield the transpose of **A**. The transpose of **A** is denoted by $\mathbf{A}^T$.

If all non-diagonal elements of **A** are zero (i.e., $a_{n,m} = 0, \quad n \neq m$), **A** is called a diagonal matrix.

If all elements of a square diagonal matrix **A** are unity, **A** is an identity matrix. Throughout the text, letter **I** is used for identity matrices.

Transpose of product of matrices **A** and **B**:

$$(\mathbf{A} \cdot \mathbf{B})^T = \mathbf{B}^T \cdot \mathbf{A}^T \tag{B-1}$$

$\mathbf{A}^{-1}$, $\mathbf{B}^{-1}$, ... are the inverses of nonsingular square matrix **A**, **B**, ...:

$$\mathbf{A} \cdot \mathbf{A}^{-1} = \mathbf{A}^{-1} \cdot \mathbf{A} = \mathbf{I} \tag{B-2}$$

Product of matrix inverses:

$$(\mathbf{A} \cdot \mathbf{B})^{-1} = \mathbf{B}^{-1} \cdot \mathbf{A}^{-1} \tag{B-3}$$

Symmetric matrix:

$$\mathbf{A}^T = \mathbf{A} \tag{B-4}$$

Skew-symmetric matrix:

$$\mathbf{A}^T = -\mathbf{A} \tag{B-5}$$

Orthogonal matrix:

$$\mathbf{A}^T = \mathbf{A}^{-1} \tag{B-6}$$

Differentiation of matrix $\mathbf{A}(\alpha)$ with respect to scalar variable α:

$$\frac{d\mathbf{A}}{d\alpha} = \begin{pmatrix} \frac{da_{1,1}}{d\alpha} & \cdots & \frac{da_{M,1}}{d\alpha} \\ \vdots & \ddots & \vdots \\ \frac{da_{N,1}}{d\alpha} & \cdots & \frac{da_{N,M}}{d\alpha} \end{pmatrix} \tag{B-7}$$

Similarity transform:

Two square matrices **A** and **B** of same size (N by N) are similar, if there exists a nonsingular matrix **C** such that

$$\mathbf{B} = \mathbf{C}^{-1} \cdot \mathbf{A} \cdot \mathbf{C} \tag{B-8}$$

Eigenvalues λ_n and associated eigenvectors $\mathbf{x}_n$ of square N-by-N matrix $\mathbf{A}$:

$$(\mathbf{A}-\lambda_n\mathbf{I})\cdot\mathbf{x}_n=\mathbf{0},\quad n=1,2,\ldots,N$$

nonzero solutions (eigenvalues) determined (B-9)

by $det(\mathbf{A}-\lambda_n\mathbf{I})=0$

Characteristic Nth-degree polynomial of N-by-N matrix $\mathbf{A}$:

$$det(\mathbf{A}-\lambda\mathbf{I}) \tag{B-10}$$

Diagonal matrix of eigenvalues:

$$\mathbf{\Lambda}=\begin{pmatrix}\lambda_1 & 0 & \cdots & 0\\ 0 & \lambda_2 & \ddots & 0\\ \vdots & \ddots & \ddots & \vdots\\ 0 & 0 & \cdots & \lambda_N\end{pmatrix} \tag{B-11}$$

Using a symmetric and orthogonal transformation matrix $\mathbf{C}$, a real symmetric matrix $\mathbf{A}$ can always be transformed into a diagonal matrix of its eigenvalues:

$$\mathbf{C}^T\cdot\mathbf{A}\cdot\mathbf{C}=\mathbf{C}^{-1}\cdot\mathbf{A}\cdot\mathbf{C}=\mathbf{\Lambda} \tag{B-12}$$

Matrix exponential:

$$exp(\mathbf{A})=\mathbf{I}+\sum_{k=1}^{\infty}\frac{\mathbf{A}^k}{k!} \tag{B-13}$$

Hermetian (= conjugate transpose) of complex matrix **A**:

$$\mathbf{A} = \begin{pmatrix} \underline{a}_{1,1} & \underline{a}_{1,2} & \cdots & \underline{a}_{1,M} \\ \underline{a}_{2,1} & \underline{a}_{2,2} & \cdots & \underline{a}_{2,M} \\ \vdots & \vdots & \ddots & \vdots \\ \underline{a}_{N,1} & \underline{a}_{N,2} & \cdots & \underline{a}_{N,M} \end{pmatrix}$$

$$\Updownarrow \tag{B-14}$$

$$\mathbf{A}^H = \begin{pmatrix} \underline{a}^*_{1,1} & \underline{a}^*_{2,1} & \cdots & \underline{a}^*_{N,1} \\ \underline{a}^*_{1,2} & \underline{a}^*_{2,2} & \cdots & \underline{a}^*_{N,2} \\ \vdots & \vdots & \ddots & \vdots \\ \underline{a}^*_{M,1} & \underline{a}^*_{M,2} & \cdots & \underline{a}^*_{M,N} \end{pmatrix}$$

Properties of Hermetian:

$$(\mathbf{A}^H)^{-1} = (\mathbf{A}^{-1})^H \tag{B-15}$$

$$(\mathbf{A} + \mathbf{B})^H = \mathbf{A}^H + \mathbf{B}^H \tag{B-16}$$

$$(\mathbf{A} \cdot \mathbf{B})^H = \mathbf{B}^H \cdot \mathbf{A}^H \tag{B-17}$$

Frobenius (or Euclidean) norm of $N \times N$ matrix **A**:

$$\|\mathbf{A}\|_F = \sqrt{\sum_{i=1}^{N} \sum_{j=1}^{N} \left|\underline{a}_{i,j}\right|^2} = \sqrt{tr(\mathbf{A}^{\mathrm{H}} \cdot \mathbf{A})} \tag{B-18}$$

where the trace of **A**, *tr*(**A**), is the sum of all diagonal elements of matrix **A**.

APPENDIX C

LIST OF SCALED-ORTHOGONAL SPACE-TIME CODE (STC) MATRICES

A space-time encoder with K input symbols and ℓ transmit periods is characterized by its **space-time code rate** (r_{STC}) given by the ratio

$$r_{STC} = \frac{K}{\ell}. \tag{C-1}$$

A number of K real or complex symbols is transmitted via N antennas over ℓ symbol time intervals.

For **real-valued** symbols (superscript "r") and **complex-valued** symbols (superscript "c") code matrices are listed below.

2 Antennas:

$$\mathbf{X}_2^{(r)} = \begin{pmatrix} x_1 & x_2 \\ -x_2 & x_1 \end{pmatrix} \tag{C-2}$$

$$\mathbf{X}_2^{(c)} = \begin{pmatrix} \underline{x}_1 & \underline{x}_2 \\ -\underline{x}_2^* & \underline{x}_1^* \end{pmatrix} \quad \text{(see example of Alamouti scheme)} \tag{C-3}$$

3 Antennas:

$$\mathbf{X}_3^{(r)} = \begin{pmatrix} x_1 & x_2 & x_3 \\ -x_2 & x_1 & -x_4 \\ -x_3 & x_4 & x_1 \\ -x_4 & -x_3 & x_2 \end{pmatrix} \tag{C-4}$$

$$\mathbf{X}_3^{(c)} = \begin{pmatrix} \underline{x}_1 & \underline{x}_2 & \underline{x}_3 \\ -\underline{x}_2 & \underline{x}_1 & -\underline{x}_4 \\ -\underline{x}_3 & \underline{x}_4 & \underline{x}_1 \\ -\underline{x}_4 & -\underline{x}_3 & \underline{x}_2 \\ \underline{x}_1^* & \underline{x}_2^* & \underline{x}_3^* \\ -\underline{x}_2^* & \underline{x}_1^* & -\underline{x}_4^* \\ -\underline{x}_3^* & \underline{x}_4^* & \underline{x}_1^* \\ -\underline{x}_4^* & -\underline{x}_3^* & \underline{x}_2^* \end{pmatrix} \quad \text{Note: } r_{STC} = 1/2 \qquad \text{(C-5)}$$

Here, blocks of $K = 4$ symbols are taken and transmitted in parallel using N=3 antennas over $\ell = 8$ symbol periods. Therefore, the code rate is ½.

With the following two codes, blocks of $K = 3$ symbols are taken and transmitted in parallel using $N = 3$ antennas over $\ell = 4$ symbol periods. Therefore, the code rate is $r_{STC} = 3/4$.

$$\mathbf{X}_3^{(c)} = \begin{pmatrix} \underline{x}_1 & \underline{x}_2 & \frac{1}{2}\sqrt{2}\underline{x}_3 \\ -\underline{x}_2^* & \underline{x}_1^* & \frac{1}{2}\sqrt{2}\underline{x}_3 \\ \frac{1}{2}\sqrt{2}\underline{x}_3 & \frac{1}{2}\sqrt{2}\underline{x}_3 & \frac{1}{2}(-\underline{x}_1 - \underline{x}_1^* + \underline{x}_2 - \underline{x}_2^*) \\ \frac{1}{2}\sqrt{2}\underline{x}_3 & -\frac{1}{2}\sqrt{2}\underline{x}_3 & \frac{1}{2}(\underline{x}_1 - \underline{x}_1^* + \underline{x}_2 + \underline{x}_2^*) \end{pmatrix} \qquad \text{(C-6)}$$

$$\mathbf{X}_3^{(c)} = \begin{pmatrix} \underline{x}_1 & -\underline{x}_2 & -\underline{x}_3 \\ \underline{x}_2^* & \underline{x}_1^* & 0 \\ \underline{x}_3^* & 0 & \underline{x}_1^* \\ 0 & -\underline{x}_3^* & \underline{x}_2^* \end{pmatrix} \quad \text{Note: } r_{STC} = 3/4. \qquad \text{(C-7)}$$

4 Antennas:

$$\mathbf{X}_4^{(r)} = \begin{pmatrix} x_1 & x_2 & x_3 & x_4 \\ -x_2 & x_1 & -x_4 & x_3 \\ -x_3 & x_4 & x_1 & -x_2 \\ -x_4 & -x_3 & x_2 & x_1 \end{pmatrix} \qquad \text{(C-8)}$$

$$\mathbf{X}_4^{(c)} = \begin{pmatrix} \underline{x}_1 & \underline{x}_2 & \underline{x}_3 & \underline{x}_4 \\ -\underline{x}_2 & \underline{x}_1 & -\underline{x}_4 & \underline{x}_3 \\ -\underline{x}_3 & \underline{x}_4 & \underline{x}_1 & -\underline{x}_2 \\ -\underline{x}_4 & -\underline{x}_3 & \underline{x}_2 & \underline{x}_1 \\ \underline{x}_1^* & \underline{x}_2^* & \underline{x}_3^* & \underline{x}_4^* \\ -\underline{x}_2^* & \underline{x}_1^* & -\underline{x}_4^* & \underline{x}_3^* \\ -\underline{x}_3^* & \underline{x}_4^* & \underline{x}_1^* & -\underline{x}_2^* \\ -\underline{x}_4^* & -\underline{x}_3^* & \underline{x}_2^* & \underline{x}_1^* \end{pmatrix} \quad \text{Note: } r_{STC} = 1/2 \qquad \text{(C-9)}$$

$$\mathbf{X}_4^{(c)} = \begin{pmatrix} \underline{x}_1 & \underline{x}_2 & \frac{1}{2}\sqrt{2}\underline{x}_3 & \frac{1}{2}\sqrt{2}\underline{x}_3 \\ -\underline{x}_2 & \underline{x}_1 & \frac{1}{2}\sqrt{2}\underline{x}_3 & \frac{1}{2}\sqrt{2}\underline{x}_3 \\ \frac{1}{2}\sqrt{2}\underline{x}_3^* & \frac{1}{2}\sqrt{2}\underline{x}_3^* & \frac{1}{2}(-\underline{x}_1 - \underline{x}_1^* + \underline{x}_2 - \underline{x}_2^*) & \frac{1}{2}(\underline{x}_1 - \underline{x}_1^* - \underline{x}_2 - \underline{x}_2^*) \\ \frac{1}{2}\sqrt{2}\underline{x}_3^* & -\frac{1}{2}\sqrt{2}\underline{x}_3^* & \frac{1}{2}(\underline{x}_1 - \underline{x}_1^* + \underline{x}_2 + \underline{x}_2^*) & \frac{1}{2}(-\underline{x}_1 - \underline{x}_1^* - \underline{x}_2 + \underline{x}_2^*) \end{pmatrix}$$

Note: $r_{STC} = 3/4$ (C-10)

More block STC matrix designs in:

[C1] H.-F. Lu, P. V. Kumar, and H. Chung. "On Orthogonal Designs and Space-Time Codes," *IEEE Communications Letters*, Vol. 8, No. 4, April 2004, pp. 220 - 222.

[C2] W. Zhao, G. Leus, and G. B. Giannakis. "Orthogonal Design of Unitary Constellations and Trellis-Coded Noncoherent Space-Time Systems," *IEEE Transactions on Information Theory*, Vol. 50, No. 6, June 2004, pp. 1319 - 1327.

INDEX